Pharmaceutical Extrusion Technology

DRUGS AND THE PHARMACEUTICAL SCIENCES

DRUGS AND THE PHARMACEUTICAL SCIENCES

A Series of Textbooks and Monographs

1. Pharmacokinetics, *Milo Gibaldi and Donald Perrier*
2. Good Manufacturing Practices for Pharmaceuticals: A Plan for Total Quality Control, *Sidney H. Willig, Murray M. Tuckerman, and William S. Hitchings IV*
3. Microencapsulation, *edited by J. R. Nixon*
4. Drug Metabolism: Chemical and Biochemical Aspects, *Bernard Testa and Peter Jenner*
5. New Drugs: Discovery and Development, *edited by Alan A. Rubin*
6. Sustained and Controlled Release Drug Delivery Systems, *edited by Joseph R. Robinson*
7. Modern Pharmaceutics, *edited by Gilbert S. Banker and Christopher T. Rhodes*
8. Prescription Drugs in Short Supply: Case Histories, *Michael A. Schwartz*
9. Activated Charcoal: Antidotal and Other Medical Uses, *David O. Cooney*
10. Concepts in Drug Metabolism (in two parts), *edited by Peter Jenner and Bernard Testa*
11. Pharmaceutical Analysis: Modern Methods (in two parts), *edited by James W. Munson*
12. Techniques of Solubilization of Drugs, *edited by Samuel H. Yalkowsky*
13. Orphan Drugs, *edited by Fred E. Karch*
14. Novel Drug Delivery Systems: Fundamentals, Developmental Concepts, Biomedical Assessments, *Yie W. Chien*
15. Pharmacokinetics: Second Edition, Revised and Expanded, *Milo Gibaldi and Donald Perrier*
16. Good Manufacturing Practices for Pharmaceuticals: A Plan for Total Quality Control, Second Edition, Revised and Expanded, *Sidney H. Willig, Murray M. Tuckerman, and William S. Hitchings IV*
17. Formulation of Veterinary Dosage Forms, *edited by Jack Blodinger*
18. Dermatological Formulations: Percutaneous Absorption, *Brian W. Barry*
19. The Clinical Research Process in the Pharmaceutical Industry, *edited by Gary M. Matoren*
20. Microencapsulation and Related Drug Processes, *Patrick B. Deasy*
21. Drugs and Nutrients: The Interactive Effects, *edited by Daphne A. Roe and T. Colin Campbell*
22. Biotechnology of Industrial Antibiotics, *Erick J. Vandamme*

23. Pharmaceutical Process Validation, *edited by Bernard T. Loftus and Robert A. Nash*
24. Anticancer and Interferon Agents: Synthesis and Properties, *edited by Raphael M. Ottenbrite and George B. Butler*
25. Pharmaceutical Statistics: Practical and Clinical Applications, *Sanford Bolton*
26. Drug Dynamics for Analytical, Clinical, and Biological Chemists, *Benjamin J. Gudzinowicz, Burrows T. Younkin, Jr., and Michael J. Gudzinowicz*
27. Modern Analysis of Antibiotics, *edited by Adjoran Aszalos*
28. Solubility and Related Properties, *Kenneth C. James*
29. Controlled Drug Delivery: Fundamentals and Applications, Second Edition, Revised and Expanded, *edited by Joseph R. Robinson and Vincent H. Lee*
30. New Drug Approval Process: Clinical and Regulatory Management, *edited by Richard A. Guarino*
31. Transdermal Controlled Systemic Medications, *edited by Yie W. Chien*
32. Drug Delivery Devices: Fundamentals and Applications, *edited by Praveen Tyle*
33. Pharmacokinetics: Regulatory • Industrial • Academic Perspectives, *edited by Peter G. Welling and Francis L. S. Tse*
34. Clinical Drug Trials and Tribulations, *edited by Allen E. Cato*
35. Transdermal Drug Delivery: Developmental Issues and Research Initiatives, *edited by Jonathan Hadgraft and Richard H. Guy*
36. Aqueous Polymeric Coatings for Pharmaceutical Dosage Forms, *edited by James W. McGinity*
37. Pharmaceutical Pelletization Technology, *edited by Isaac Ghebre-Sellassie*
38. Good Laboratory Practice Regulations, *edited by Allen F. Hirsch*
39. Nasal Systemic Drug Delivery, *Yie W. Chien, Kenneth S. E. Su, and Shyi-Feu Chang*
40. Modern Pharmaceutics: Second Edition, Revised and Expanded, *edited by Gilbert S. Banker and Christopher T. Rhodes*
41. Specialized Drug Delivery Systems: Manufacturing and Production Technology, *edited by Praveen Tyle*
42. Topical Drug Delivery Formulations, *edited by David W. Osborne and Anton H. Amann*
43. Drug Stability: Principles and Practices, *Jens T. Carstensen*
44. Pharmaceutical Statistics: Practical and Clinical Applications, Second Edition, Revised and Expanded, *Sanford Bolton*
45. Biodegradable Polymers as Drug Delivery Systems, *edited by Mark Chasin and Robert Langer*
46. Preclinical Drug Disposition: A Laboratory Handbook, *Francis L. S. Tse and James J. Jaffe*
47. HPLC in the Pharmaceutical Industry, *edited by Godwin W. Fong and Stanley K. Lam*
48. Pharmaceutical Bioequivalence, *edited by Peter G. Welling, Francis L. S. Tse, and Shrikant V. Dinghe*

49. Pharmaceutical Dissolution Testing, *Umesh V. Banakar*
50. Novel Drug Delivery Systems: Second Edition, Revised and Expanded, *Yie W. Chien*
51. Managing the Clinical Drug Development Process, *David M. Cocchetto and Ronald V. Nardi*
52. Good Manufacturing Practices for Pharmaceuticals: A Plan for Total Quality Control, Third Edition, *edited by Sidney H. Willig and James R. Stoker*
53. Prodrugs: Topical and Ocular Drug Delivery, *edited by Kenneth B. Sloan*
54. Pharmaceutical Inhalation Aerosol Technology, *edited by Anthony J. Hickey*
55. Radiopharmaceuticals: Chemistry and Pharmacology, *edited by Adrian D. Nunn*
56. New Drug Approval Process: Second Edition, Revised and Expanded, *edited by Richard A. Guarino*
57. Pharmaceutical Process Validation: Second Edition, Revised and Expanded, *edited by Ira R. Berry and Robert A. Nash*
58. Ophthalmic Drug Delivery Systems, *edited by Ashim K. Mitra*
59. Pharmaceutical Skin Penetration Enhancement, *edited by Kenneth A. Walters and Jonathan Hadgraft*
60. Colonic Drug Absorption and Metabolism, *edited by Peter R. Bieck*
61. Pharmaceutical Particulate Carriers: Therapeutic Applications, *edited by Alain Rolland*
62. Drug Permeation Enhancement: Theory and Applications, *edited by Dean S. Hsieh*
63. Glycopeptide Antibiotics, *edited by Ramakrishnan Nagarajan*
64. Achieving Sterility in Medical and Pharmaceutical Products, *Nigel A. Halls*
65. Multiparticulate Oral Drug Delivery, *edited by Isaac Ghebre-Sellassie*
66. Colloidal Drug Delivery Systems, *edited by Jörg Kreuter*
67. Pharmacokinetics: Regulatory • Industrial • Academic Perspectives, Second Edition, *edited by Peter G. Welling and Francis L. S. Tse*
68. Drug Stability: Principles and Practices, Second Edition, Revised and Expanded, *Jens T. Carstensen*
69. Good Laboratory Practice Regulations: Second Edition, Revised and Expanded, *edited by Sandy Weinberg*
70. Physical Characterization of Pharmaceutical Solids, *edited by Harry G. Brittain*
71. Pharmaceutical Powder Compaction Technology, *edited by Göran Alderborn and Christer Nyström*
72. Modern Pharmaceutics: Third Edition, Revised and Expanded, *edited by Gilbert S. Banker and Christopher T. Rhodes*
73. Microencapsulation: Methods and Industrial Applications, *edited by Simon Benita*
74. Oral Mucosal Drug Delivery, *edited by Michael J. Rathbone*
75. Clinical Research in Pharmaceutical Development, *edited by Barry Bleidt and Michael Montagne*

76. The Drug Development Process: Increasing Efficiency and Cost Effectiveness, *edited by Peter G. Welling, Louis Lasagna, and Umesh V. Banakar*

77. Microparticulate Systems for the Delivery of Proteins and Vaccines, *edited by Smadar Cohen and Howard Bernstein*

78. Good Manufacturing Practices for Pharmaceuticals: A Plan for Total Quality Control, Fourth Edition, Revised and Expanded, *Sidney H. Willig and James R. Stoker*

79. Aqueous Polymeric Coatings for Pharmaceutical Dosage Forms: Second Edition, Revised and Expanded, *edited by James W. McGinity*

80. Pharmaceutical Statistics: Practical and Clinical Applications, Third Edition, *Sanford Bolton*

81. Handbook of Pharmaceutical Granulation Technology, *edited by Dilip M. Parikh*

82. Biotechnology of Antibiotics: Second Edition, Revised and Expanded, *edited by William R. Strohl*

83. Mechanisms of Transdermal Drug Delivery, *edited by Russell O. Potts and Richard H. Guy*

84. Pharmaceutical Enzymes, *edited by Albert Lauwers and Simon Scharpé*

85. Development of Biopharmaceutical Parenteral Dosage Forms, *edited by John A. Bontempo*

86. Pharmaceutical Project Management, *edited by Tony Kennedy*

87. Drug Products for Clinical Trials: An International Guide to Formulation • Production • Quality Control, *edited by Donald C. Monkhouse and Christopher T. Rhodes*

88. Development and Formulation of Veterinary Dosage Forms: Second Edition, Revised and Expanded, *edited by Gregory E. Hardee and J. Desmond Baggot*

89. Receptor-Based Drug Design, *edited by Paul Leff*

90. Automation and Validation of Information in Pharmaceutical Processing, *edited by Joseph F. deSpautz*

91. Dermal Absorption and Toxicity Assessment, *edited by Michael S. Roberts and Kenneth A. Walters*

92. Pharmaceutical Experimental Design, *Gareth A. Lewis, Didier Mathieu, and Roger Phan-Tan-Luu*

93. Preparing for FDA Pre-Approval Inspections, *edited by Martin D. Hynes III*

94. Pharmaceutical Excipients: Characterization by IR, Raman, and NMR Spectroscopy, *David E. Bugay and W. Paul Findlay*

95. Polymorphism in Pharmaceutical Solids, *edited by Harry G. Brittain*

96. Freeze-Drying/Lyophilization of Pharmaceutical and Biological Products, *edited by Louis Rey and Joan C. May*

97. Percutaneous Absorption: Drugs–Cosmetics–Mechanisms–Methodology, Third Edition, Revised and Expanded, *edited by Robert L. Bronaugh and Howard I. Maibach*

98. Bioadhesive Drug Delivery Systems: Fundamentals, Novel Approaches, and Development, *edited by Edith Mathiowitz, Donald E. Chickering III, and Claus-Michael Lehr*
99. Protein Formulation and Delivery, *edited by Eugene J. McNally*
100. New Drug Approval Process: Third Edition, The Global Challenge, *edited by Richard A. Guarino*
101. Peptide and Protein Drug Analysis, *edited by Ronald E. Reid*
102. Transport Processes in Pharmaceutical Systems, *edited by Gordon L. Amidon, Ping I. Lee, and Elizabeth M. Topp*
103. Excipient Toxicity and Safety, *edited by Myra L. Weiner and Lois A. Kotkoskie*
104. The Clinical Audit in Pharmaceutical Development, *edited by Michael R. Hamrell*
105. Pharmaceutical Emulsions and Suspensions, *edited by Francoise Nielloud and Gilberte Marti-Mestres*
106. Oral Drug Absorption: Prediction and Assessment, *edited by Jennifer B. Dressman and Hans Lennernäs*
107. Drug Stability: Principles and Practices, Third Edition, Revised and Expanded, *edited by Jens T. Carstensen and C. T. Rhodes*
108. Containment in the Pharmaceutical Industry, *edited by James P. Wood*
109. Good Manufacturing Practices for Pharmaceuticals: A Plan for Total Quality Control from Manufacturer to Consumer, Fifth Edition, Revised and Expanded, *Sidney H. Willig*
110. Advanced Pharmaceutical Solids, *Jens T. Carstensen*
111. Endotoxins: Pyrogens, LAL Testing, and Depyrogenation, Second Edition, Revised and Expanded, *Kevin L. Williams*
112. Pharmaceutical Process Engineering, *Anthony J. Hickey and David Ganderton*
113. Pharmacogenomics, *edited by Werner Kalow, Urs A. Meyer, and Rachel F. Tyndale*
114. Handbook of Drug Screening, *edited by Ramakrishna Seethala and Prabhavathi B. Fernandes*
115. Drug Targeting Technology: Physical • Chemical • Biological Methods, *edited by Hans Schreier*
116. Drug–Drug Interactions, *edited by A. David Rodrigues*
117. Handbook of Pharmaceutical Analysis, *edited by Lena Ohannesian and Anthony J. Streeter*
118. Pharmaceutical Process Scale-Up, *edited by Michael Levin*
119. Dermatological and Transdermal Formulations, *edited by Kenneth A. Walters*
120. Clinical Drug Trials and Tribulations: Second Edition, Revised and Expanded, *edited by Allen Cato, Lynda Sutton, and Allen Cato III*
121. Modern Pharmaceutics: Fourth Edition, Revised and Expanded, *edited by Gilbert S. Banker and Christopher T. Rhodes*
122. Surfactants and Polymers in Drug Delivery, *Martin Malmsten*
123. Transdermal Drug Delivery: Second Edition, Revised and Expanded, *edited by Richard H. Guy and Jonathan Hadgraft*

124. Good Laboratory Practice Regulations: Second Edition, Revised and Expanded, *edited by Sandy Weinberg*

125. Parenteral Quality Control: Sterility, Pyrogen, Particulate, and Package Integrity Testing: Third Edition, Revised and Expanded, *Michael J. Akers, Daniel S. Larrimore, and Dana Morton Guazzo*

126. Modified-Release Drug Delivery Technology, *edited by Michael J. Rathbone, Jonathan Hadgraft, and Michael S. Roberts*

127. Simulation for Designing Clinical Trials: A Pharmacokinetic-Pharmacodynamic Modeling Perspective, *edited by Hui C. Kimko and Stephen B. Duffull*

128. Affinity Capillary Electrophoresis in Pharmaceutics and Biopharmaceutics, *edited by Reinhard H. H. Neubert and Hans-Hermann Rüttinger*

129. Pharmaceutical Process Validation: An International Third Edition, Revised and Expanded, *edited by Robert A. Nash and Alfred H. Wachter*

130. Ophthalmic Drug Delivery Systems: Second Edition, Revised and Expanded, *edited by Ashim K. Mitra*

131. Pharmaceutical Gene Delivery Systems, *edited by Alain Rolland and Sean M. Sullivan*

132. Biomarkers in Clinical Drug Development, *edited by John C. Bloom and Robert A. Dean*

133. Pharmaceutical Extrusion Technology, *edited by Isaac Ghebre-Sellassie and Charles Martin*

ADDITIONAL VOLUMES IN PREPARATION

Pharmaceutical Inhalation Aerosol Technology: Second Edition, Revised and Expanded, *edited by Anthony J. Hickey*

Pharmaceutical Compliance, *edited by Carmen Medina*

Pharmaceutical Statistics: Practical and Clinical Applications, Fourth Edition, *Sanford Bolton and Charles Bon*

Pharmaceutical Extrusion Technology

edited by
Isaac Ghebre-Sellassie
MEGA Pharmaceuticals
Asmara, Eritrea
and Pharmaceutical Technology Solutions
Morris Plains, New Jersey, U.S.A.

Charles Martin
American Leistritz Extruder Corporation
Somerville, New Jersey, U.S.A.

MARCEL DEKKER, INC.　　　　　NEW YORK • BASEL

Library of Congress Cataloging-in-Publication Data
A catalog record for this book is available from the Library of Congress.

ISBN: 0-8247-4050-5

This book is printed on acid-free paper.

Headquarters
Marcel Dekker, Inc., 270 Madison Avenue, New York, NY 10016, U.S.A.
tel: 212-696-9000; fax: 212-685-4540

Distribution and Customer Service
Marcel Dekker, Inc., Cimarron Road, Monticello, New York 12701, U.S.A.
tel: 800-228-1160; fax: 845-796-1772

Eastern Hemisphere Distribution
Marcel Dekker AG, Hutgasse 4, Postfach 812, Ch-4001 Basel, Switzerland
tel: 41-61-260-6300; fax: 41-61-260-6333

World Wide Web
http://www.dekker.com

The publisher offers discounts on this book when ordered in bulk quantities. For more information, write to Special Sales/Professional Marketing at the headquarters address above.

Current printing (last digit):
10 9 8 7 6 5 4 3 2 1

PRINTED IN THE UNITED STATES OF AMERICA

Preface

Extrusion is a process in which materials are mixed intimately under controlled conditions of temperature, shear, and pressure to generate a variety of in-process and finished products using pieces of equipment collectively known as extruders. Most of the early applications of this technology were limited to industrial processes that ranged from compaction of clay and straw to the mastication of rubber and plastics. Prominent in its absence throughout this period was the application of extrusion in the development and manufacture of pharmaceutical products. As a result, until recently, the pharmaceutical industry contributed very little to the technological advances made either in the improvement of existing extrusion equipment or in the design of specialized extruders that satisfy its highly regulated sphere of operation. However, as soon as work started on the applicability of the technology in pharmaceutical dosage from development, it became apparent that extrusion processes not only improve the efficiency of pharmaceutical manufacturing processes, but also can significantly enhance the quality of manufactured products owing to the mixing efficiency of extruders. The intent of this book is, therefore, to provide pharmaceutical scientists and technologies with basic principles and fundamentals of extrusion technology and a detailed description of the practical applications of extrusion processes.

Chapter 1 provides a historical overview of extruders and extrusion technology, and describes the evolutionary development of the technology in an attempt to address the processing needs of the time. Early patents that formed the basis for the widespread use of the technology are highlighted. Extrusion design features that are common to general industrial applications, but also uniquely applicable to the pharmaceutical industry, including cleanability of the product contact parts, are discussed in Chapter 2. Single-crew extruders, the first family of extruders to be widely introduced into the market, are continuous, high-pressure generating pumps that also perform limited mixing and devolatilization functions. These extruders, as well as the impact of different system configurations on the processability of various formulations, are extensively discussed in Chapter 3. Chapter 4 addresses twin-screw extruders. Corotating and counterrotating twin-screw compounding extruders are mass transfer devices that are used to mix together two or more materials into a homogeneous mass in a continuous process. This is accomplished through distributive and dispersive mixing of the various components in a formulation. These aspects as well as screw designs specific to twin-screw extruders are discussed in depth.

Extrusion dies, located at the exit end of the barrel assembly, are critical parts of the extrusion system that can literally make or break a process, particularly during the manufacture of controlled-release dosage forms. Success of a given process depends on, among other factors, die design. While some dies may be simple, others can be extremely complex. These and some of the techniques employed in die design are reviewed in Chapter 5. Chapter 6 covers material handling and feeder technology. During extrusion, liquids and powders must be fed accurately in a continuous manner throughout the run time, including the refill period, to ensure formulation consistency, constant throughput, proper order of mixing of ingredients, and regulated mass transfer. Feeders are hence critical components of the extrusion line that need to be evaluated carefully. A detailed description of torque rheometers, instruments that are essential for the characterization of formulations and processing parameters, and techniques that help elucidate the rheological properties of melts, are presented in Chapter 7.

Chapter 8 describes general extrusion processes and troubleshooting, and delineates the impact each process parameter has on the characteristics of the overall manufacturing process. Chapter 9 covers melt pelletization processes, and provides a review of the various pelletizers that have been in use since the 1950s. Chapter 10 focuses specifically on critical formulation parameters that affect the development of melt-extruded controlled release pellets, with particular emphasis on the effect of various formulation components on the release profiles of oral solid dosage forms. Shape extrusion involving solid or hollow structures is discussed in Chapter 11. Chapter 12 describes in detail technologies relevant to the manufacturing of sheets and laminates, materials that could have extensive application in transdermal drug delivery and fast-dissolving films intended for oral drug delivery. The enhancement of dissolution rates, and hence bioavailabilities, of poorly water-soluble drug substances formulated as melt-extruded molecular and particulate dispersions is discussed in Chapters 13 and 14.

Chapter 15 covers recent advances made in the area of extrusion/spheronization, a popular solvent-based pelletization process that is employed, in the majority of cases, to manufacture high-potency pellets. Key process variables and formulation factors that determine the quality of pellets are highlighted. Wet granulation processes, which traditionally have employed planetary and high-shear mixers, can also be carried out using modified twin-screw extruders. Extruders, being high-intensity, small-volume mixers, produce very uniform granulations that allow the manufacture of precisely designed oral dosage forms and are discussed in chapter 16. Control systems and instrumentation, which provide efficient mechanisms to control and monitor key process variables and data-acquisition systems, are discussed in Chapter 17. Chapter 18 provides an overview of the installation, commissioning, and qualification requirements for extruders employed in the pharmaceutical industry. Given the advantages that extrusion technology offers, interest in the technology is expected to grow, thereby challenging equipment manufacturers and process engineers to further refine and expand the application of the technology in the development and manufacturing of pharmaceutical dosage forms, with particular emphasis on equipment design and configurations. Some of these potential future developments and benefits are summarized in Chapter 19.

This book is the first of its kind that discusses extensively the well-developed science of extrusion technology as applied to pharmaceutical drug product development and manufacturing. By covering a wide range of relevant topics, the text brings together all technical information necessary to develop and market pharmaceutical dosage forms that meet current quality and regulatory requirements. As extrusion technology continues to be refined further, usage of extruder systems and the array of applications will continue to expand, but the core technologies as represented in *Pharmaceutical Extrusion Technology* will remain the same.

Isaac Ghebre-Sellassie
Charles Martin

Contents

Preface   *iii*

Contributors   *ix*

1. Historical Overview   1
   Matthew Mollan

2. Extruder Design   19
   Richard Steiner

3. Single-Screw Extrusion and Screw Design   39
   Keith Luker

4. Twin-Screw Extrusion and Screw Design   69
   William Thiele

5. Die Design   99
   John Perdikoulias and Tom Dobbie

6. Material Handling and Feeder Technology   111
   Winfried Doetsch

7. Rheology and Torque Rheometers   135
   Scott T. Martin

8. Process Design   153
   Adam Dreiblatt

9. Melt Pelletization   171
   Christopher C. Case

10. Melt-Extruded Controlled-Release Dosage Forms   183
    James W. McGinity and Feng Zhang

11. Shape Extrusion   209
    Bob Bessemer

12. Film, Sheet, and Laminates   225
    Bert Elliott

13. Melt-Extruded Molecular Dispersions   245
    Jörg Breitenbach and Markus Mägerlein

14. Melt-Extruded Particulate Dispersions   261
    S. Craig Dyar, Matthew Mollan,
    and Isaac Ghebre-Sellassie

15. Extrusion/Spheronization   277
    David F. Erkoboni

16. Twin-Screw Wet Granulation   323
    Mayur Lodaya, Matthew Mollan,
    and Isaac Ghebre-Sellassie

17. Installation, Commissioning, and Qualification   345
    Adam Dreiblatt

18. Controls and Instrumentation   361
    Stuart J. Kapp and Pete A. Palmer

19. Future Trends   383
    Isaac Ghebre-Sellassie and Charles Martin

Index   *393*

Contributors

Bob Bessemer   The Conair Group, Inc., Pittsburgh, Pennsylvania, USA

Jörg Breitenbach, Ph.D.   Knoll Soliqs, Abbott GmbH and Company, Ludwigshafen, Germany

Christopher C. Case   Conair Reduction Engineering, Pittsburgh, Pennsylvania, USA

Tom Dobbie, Ph.D.   Porpoise Viscometers Ltd., Lancashire, England

Winfried Doetsch, Ph.D.   K-Tron Switzerland Ltd., Niederlenz, Switzerland

Adam Dreiblatt   Extrusioneering International, Inc., Randolph, New Jersey, USA

S. Craig Dyar, Ph.D.   Division of Pharmaceutical Research and Development, Pfizer, Inc., Ann Arbor, Michigan, USA

Bert Elliott   American Leistritz Extruder Corporation, Somerville, New Jersey, USA

David F. Erkoboni, Ph.D.   BioPolymer Division, FMC Corporation, Princeton, New Jersey, USA

Isaac Ghebre-Sellassie, Ph.D.   MEGA Pharmaceuticals, Asmara, Eritrea, and Pharmaceutical Technology Solutions, Morris Plains, New Jersey, USA

Stuart J. Kapp American Leistritz Extruder Corporation, Somerville, New Jersey, USA

Mayur Lodaya, Ph.D. Division of Pharmaceutical Research and Development, Pfizer, Inc., Ann Arbor, Michigan, USA

Keith Luker Randcastle Extrusion Systems, Cedar Grove, New Jersey, USA

Markus Mägerlein, Ph.D. Knoll Soliqs, Abbott GmbH and Company, Ludwigshafen, Germany

Charles Martin American Leistritz Extruder Corporation, Somerville, New Jersey, USA

Scott T. Martin Thermo Electron Material Characterization, Madison, Wisconsin, USA

James W. McGinity, Ph.D. College of Pharmacy, The University of Texas at Austin, Austin, Texas, USA

Matthew Mollan, Ph.D. Division of Pharmaceutical Research and Development, Pfizer, Inc., Ann Arbor, Michigan, USA

Pete A. Palmer Wolock & Lott Transmission Equipment Corporation, North Branch, New Jersey, USA

John Perdikoulias Compuplast International Inc., Ontario, Canada

Richard Steiner Leistritz Extrusionstechnik GmbH, Nuremberg, Germany

William Thiele American Leistritz Extruder Corporation, Somerville, New Jersey, USA

Feng Zhang, Ph.D. PharmaForm L.L.C., Austin, Texas, USA

Pharmaceutical
Extrusion Technology

1

Historical Overview

Matthew Mollan
Pfizer, Inc., Ann Arbor, Michigan, USA

I. INTRODUCTION

Extrusion is a well-known processing technology that has been developed over the last century and spans many diverse industrial fields. Extrusion can be simply defined as the process of forming a new material (the extrudate) by forcing a material through an orifice or die under controlled conditions. The historical use of extruders, and especially of twin-screw extruders, has been with the processing of foods and the manufacturing of plastics. Developments in multiscrew extrusion-based processing have occurred throughout the industrialized world, with the major early advances occurring in Germany, the United Kingdom, and later in the United States.

The widespread industrial use of single-screw extruders came of age both with the extrusion of thermoplastic materials in the early 1930s (1) in the polymer industry, as well as with the continuous extrusion of pasta products in 1935 (2) in the food industry. Pharmaceutical applications of ram extrusion and single-screw extrusion are well known in spheronization technologies for the formation of multiparticulates. Ram and single-screw extruders, as well as some pharmaceutical applications of those types of extruders, have been previously reviewed (3). Twin-screw extruders were initially developed in the 1800s, with the concept of combining the machine actions of several available devices into a single unit. Although both single-screw and twin-screw extruders were initially developed within a relatively similar time frame, the commercialization and the widespread use of single-screw systems occurred much earlier than that of twin-screw systems. Twin-screw extruders had several significant engineering issues that were not fully overcome until the 1940s.

The use of twin-screw extrusion processing is not very well characterized to date for pharmaceutical technology, especially with regard to melt extrusion. Melt extrusion is a process where materials are melted, and then passed through a die to form an extrudate. Melt extrusion processing for pharmaceuticals is used in the manufacture of solid dispersions and controlled release products. It is also an efficient continuous process for dosage form manufacture with the incorporation of calendering or injection molding equipment. Melt extrusion can also be used as part of the process for manufacturing implantable devices.

Twin-screw extruders have many potential applications in pharmaceutical manufacturing due to their inherent design and operating characteristics. These same design advantages had previously led to the widespread use of twin-screw extruders in other industries. It is significant to note that twin-screw extruders usage in other industries is generally with products that are much less valuable on a per-kilogram basis as compared to pharmaceuticals. In spite of the cost-per-kilogram issues in other industries, the technology rapidly proved its value, and had been considered "conventional" for many years prior to any work in the pharmaceutical industry. The advantage of twin-screw extrusion processing, which has led to its widespread industrial usage, is that a number of functions can be performed as a single unit operation. Extruders can be used for the mixing, melting, and reacting of materials, i.e., compounding within a single devise. This has the practical effect of combining a number of previously separate batch operations into a single unit operation, thus increasing manufacturing efficiency. An extruder can also be used in batch mode, but its continuous processing operating characteristics allow the use of large-scale batch sizes, i.e., batch sizes in the thousands of kilograms. Extruders also lend themselves to being used in a continuous manner, and the use of an extruder as a continuous processor is especially advantageous when attempting to minimize manufacturing plant size and costs. Several reviews of twin-screw extruders from an engineering perspective are available (4,5).

II. EXTRUDER DESCRIPTION

A basic extruder layout consists of a platform (on which is supported a drive system), barrels, screws arranged on a screw shaft, a die, and connection to utilities and controls. Extruder drive systems generally consist of a motor, a gearbox, a linkage, and thrust bearings. The overcoming of engineering design issues with thrust bearings has allowed very significant advances in twin-screw processing to rapidly occur since the 1940s.

Extruder barrels and screws are usually modular in arrangement; however, some systems are designed as a single barrel or screw. An illustration of a good manufacturing practices (GMP)-formatted twin-screw extrusion system is shown in Figure 1. An extruder is designed to support the screws in the barrel, and so must be capable of rotating the screws at the selected speed while compensating for the generated torque from the material being extruded. The torque and speed requirements of a specific formulation and the process conditions will determine the power require-ments of the extruder. The size of an extruder is generally described based

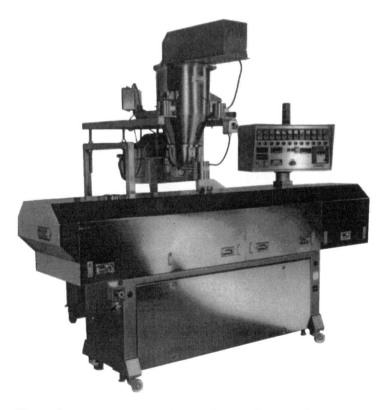

Figure 1 Twin-screw extruder in GMP format. (Courtesy of American Leistritz Co.)

on the diameter of the screw used in the system, i.e., 27 mm extruder (pilot scale) as compared with 60 mm extruder (production scale). Although the screw size terminology difference appears small ($\approx 2\times$) in the preceding example, the extruder output that results from doubling of the screw size may be 10-fold, i.e., 10–100 kg/hr. This is due to the much larger volume available for processing as the screw size is increased.

Twin-screw extruders create a controlled temperature and pressure environment in which a material is processed. This environment is created and maintained by controlling the temperature of each of the barrels/zones, while the pressure is controlled by the screw design and the operating conditions chosen for processing. Each of these zones is fitted with its own heating and cooling manifold, and is controlled by temperature sensors with a feedback loop controller. Most extruders have at least three regions: a feed zone, a mixing/heating/melting zone, and a pumping or metering zone. The melt extrusion process can then be visualized as particulate solid conveying, then melt conveying where solid and melt coexist, then melt conveying by partially filled channels, and finally melt conveying by more fully filled channels. Mathematical predictive modeling of material behavior in the twin-screw extruder is quite difficult because of the complexity that occurs due to the many engineering variables in screw design, as well as the non-Newtonian nature of most polymeric materials that are processed.

Once a twin-screw extruder is in operation, heating can occur both due to the frictional forces generated by the screw turning in the barrel when a material is present, as well as the heat generation due to the shearing of material in the extruder. As would be expected, shear-induced heating is highest at the point where the applied shear is highest. Excessive shear heating can be responsible for localized overheating of the material being processed, but can be controlled by proper screw design. For these reasons, the cooling of an extruder is usually required to remove both the barrel-supplied heat and/or the frictional/shear heat generated by the twin-screw extrusion process. The cooling will also minimize potential localized overheating and will avoid the occurrence of detrimental product characteristics.

III. SINGLE-SCREW EXTRUDERS

A. History

Single-screw extruders are the simplest of the screw extruder designs, and were invented to overcome processing difficulties that arose from using commercially available equipment in the late 1800s. One of the earliest patents on single-screw extruders is attributed to Sturges in 1871, and describes an extruder that was designed as a spirally arranged flange on a shaft or screw for the purpose of pumping soap (7). It is noteworthy that this patent does specifically claim the combination of separate unit operations into a single step, "The apparatus for cooling, conveying, and mixing soap. . ." An illustration from this patent is shown in Figure 2. Moving ahead a number of decades, a technically important patent in single-screw extruder development was attributed to Beck in 1952 involving reciprocating screw injection, with the polymer being "pumped" by the rotation of the screw. The process described involved having a modification to a barrel resulting in an increased pressure that leads to the material being pushed out of the extruder into a mold (8) and, as such, expands the capabilities of a single-screw extruder.

B. Operating Principles

Single-screw extruders work under the principle of flood feeding, in that the hopper is filled, and the screw takes what it wants. This has the effect on processing in that increases in screw speed therefore increase the output level. From an engineering perspective, a very important fact is that there exists a

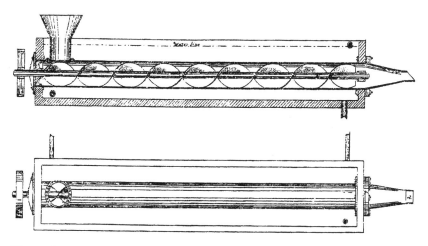

Figure 2 An illustration of single-screw extruder from Sturges. (From Ref. 7.)

continuous material bed throughout the single-screw extruder due to the screw design. The continuous bed of material then allows the transmission of pressure back from the die end to the hopper end of the extruder. The operative result of the pressure transmission is that the filling process—and therefore the actual output—is dependent on die pressure (4). The die pressure then is an indication of the pressure over a large portion of the single-screw extruder due to the presence of the continuous material bed.

C. Single-Screw and Twin-Screw Extruders

The major differences between single-screw and twin-screw extruders is in their conveying or transport mechanism (10), and in their mixing abilities. In single-screw extruders, the transport mechanism is based on frictional forces in the solids conveying zone, as well as on viscous forces in the melt conveying zone. Single-screw extrusion is therefore highly dependent on the frictional and viscous properties of the material being processed. Because material transport is dependent on frictional forces, then transport is affected by the difference in friction between the polymer and the screw, as compared with the difference in friction between the polymer and the barrel wall. An important implication is that a stagnant layer may remain at the screw surface, and the material can potentially rotate with the screw. The rotation of material with the screw obviously is to be avoided, and it can be imagined that many detrimental effects to a material can occur if a material does spend excessive amounts of time within the single-screw extruder.

A comparison of the mixing ability by the addition of a color pigment to a material being processed in a single-screw extruder and in a twin-screw extruder was shown by Ferns (9). The single-screw extruded material exhibited streaks and differences in shading, which was indicative of poor mixing. The material produced by the twin-screw extruder had no streaks and the material color was very uniform. The rationale behind these results is because mixing in a twin-screw extruder occurs both at the macroscopical level, where the material is exchanged from one screw to another, as well as at the microscopical level, where mixing occurs at the high-shear regions of screw elements interactions.

Single-screw extruders are less efficient processors than twin-screw extruders, and therefore have a generally longer equipment length. Single-screw extruders do have the advantage over twin-screw extruders in terms of their mechanical simplicity and the large cost difference. Single-screw extruders are significantly less expensive than twin-screw extruders, and so can provide high productivity-to-cost ratios. The advantages of twin-screw extruders over single-screw extruders are due to the shorter residence times of materials in the extruder, the stability of the melting process, and the smaller equipment size required to achieve an equivalent output.

IV. TWIN-SCREW EXTRUDERS
A. Background

As the name implies, twin-screw extruders use two screws usually arranged side by side. The use of two screws allows a number of different configurations to be obtained. The two screws can either rotate in the same direction (corotation), or the screws can rotate in opposite directions (counterrotation). Corotational screws can rotate either clockwise or counterclockwise, and both directions are equivalent from a processing standpoint. If the two screws rotate in different directions (either rotate toward the center, or rotate away from the center), then they are known as counterrotating extruders. Both rotational designs with twin-screw extruders have the basic processing advantages of positive conveying and effective mixing as compared with single-screw extruders. Illustrations of the two types (corotating and counterrotating) are shown in Figure 3.

The use of twin screws, instead of a single screw, imposes different conditions on all zones of the extruder, from the transfer of material from the hopper to the screw, all the way to the metered pumping zone. Material transport in a twin-screw extruder depends on the specific configuration of the screws, but, generally, the process material is prevented from rotating with the

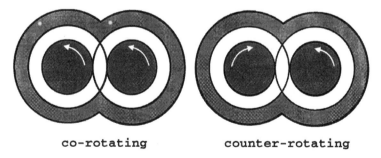

co-rotating   **counter-rotating**

Figure 3 Twin-screw configurations: corotating (left) and counterrotating (right).

screw due to the opposing screw. This is a significant advantage with twin-screw extruders over single-screw extruders that can have the detrimental processing effect of the material actually rotating with the screw, and not being conveyed forward in an efficient manner.

B. Types of Twin-Screw Extruders

There are two primary types of twin-screw extruders, counterrotating and corotating, and the differences are illustrated in Figure 3. These two primary types then can be further separated into nonintermeshing and fully intermeshing. The major difference between nonintermeshing and intermeshing twin-screw extruders is that nonintermeshing extruders cannot form closed or semiclosed compartments, and therefore have a lower degree of positive conveying characteristics. Positive conveying requires that the backflow of material is minimized, and this is best achieved by fully intermeshing screw designs. Another significant engineering design impact is that the intermeshing of the two screws has the result that no continuous bed of material exists through the length of the extruder because the material bed is broken by the flights of the screw. This is a significant difference compared to the simple designs of single-screw systems. The lack of a continuous bed through the extruder has the implications that minimal pressure transmission occurs, and the throughput, within limits of reasonable operation, should be independent of the die pressure. Several comparisons between different extruder designs have been published; however, it must be kept in mind that the specific screw design utilized has a large influence on performance. The significant impact of screw design thus makes direct comparisons between designs difficult (13,14). Additionally, many of the extruder equipment vendors also have specific differences, called "features," between their respective systems, and this also makes comparisons between equipment difficult.

C. Intermeshing and Nonintermeshing

The fully intermeshing type of screw design is the most popular type of design used for twin-screw extruders. The fully intermeshing design is self-wiping, and prevents the material from rotating with the screw. This then allows the extruder to operate by a *first in/first out* principle. The self-wiping characteristic is also beneficial as it minimizes the nonmotion of material within the extruder, thus preventing localized overheating. Fully intermeshing systems also generally have residence time distributions (the time the material is physically in the extruder), which have much sharper tails than do other extruder screw design layouts. This is a beneficial characteristic because all the materials experience an equivalent processing history. Nonintermeshing machines are often used for processing when large amounts of volatiles need to be removed from a material due to the large vent opening that can be accommodated as the screws are positioned apart from each other. They are also used when processing highly viscous materials, where the intermeshing can cause problematical torque buildups.

D. Counterrotating and Corotating Twin-Screw Extruders

Counterrotating twin-screw extruders have very positive material feed and conveying characteristics due to the material movement within the extruder. The residence time and the temperature of material in this type of twin-screw extruder are also uniform. Applications for this extruder design are used when very high shear regions are needed, and the layout is good for dispersing particles in a blend. Counterrotating twin-screw extruders subject materials to very high shear forces as the material is squeezed through the gap between the two screws as they come together. Counterrotating twin-screw extruders suffer from the disadvantages of potential air entrapment, high-pressure generation, and low maximum screw speeds and output. Counterrotating twin-screw extruders are operated at lower speeds than corotating twin-screw extruders due to the pressure that develops between the two screws. This pressure develops because the material tends to push the screws outward as the two screws come together in rotation. This outward pushing effect can also lead to barrel wall scraping by the screws with counterrotating designs, and must be monitored.

Corotating twin-screw extruders are generally of the intermeshing design, and are thus self-wiping. High screw speeds, in the hundreds of revolutions per minute or more, and high outputs are possible with corotating extruders. Corotating extruders generally experience lower screw and barrel wear than counterrotating extruders because they do not experience the outward "pushing" effect due to screw rotation. Geometrical engineering designs considerations for screw elements offer much more flexibility in the complete screw layout with corotational extrusion systems as compared to counterrotational systems. A great variety of screw elements are available, which allow the corotating twin-screw extruder to perform a multitude of tasks.

V. TWIN-SCREW EXTRUDER DESIGNS

A. Nonintermeshing Counterrotating Twin-Screw Extruders

The nonintermeshing counterrotating type of twin-screw extruder has the simplest mechanical design, and the material moves through the extruder by drag flow caused by the relative motion between the screw and the barrel. At the interscrew region, there is only pressure-driven flow. Pfleiderer (11) in 1881 patented a continuous tangential counterrotating twin-screw mixing device, and this patent formed the basis for this class of machines. The Farrel continuous mixer was patented by Ahlefeld in 1964, and involved the combination of screw elements with internal mixer rotors. The mixing elements first involved flights of forward pitch, which caused forward pumping, followed by zones with reversed pitch, which induced backward pumping. The machine was intended for continuous compounding, and was called a continuous internal stiff gel mixer (15,16). Illustrations from the patents are shown in Figures 4 and 5. These patents are also interesting because they spell out in detail that the machine described is a continuous system, and thus can be used for processing where previously only batch mode processing was possible. The engineering aspects of nonintermeshing counter-rotating twin-screw extruders have been previously examined (17).

B. Intermeshing Counterrotating Twin-Screw Extruders

The intermeshing counterrotating twin-screw extruders are derived from positive displacement twin-screw pumps, and studies in material flow visualization confirmed the positive displacement pump characteristics of the material motion in the extruder. An illustration of an intermeshing counterrotating screw is shown at the bottom example in Figure 6. An early patent in 1874 by Wiegand (18) described a fully intermeshing counterrotating extruder for dough production, and an illustration from the patent is shown in Figure 7. In 1939 and 1940, Leistritz and Burghauser described a new design of intermeshing counter-rotating twin-screw extruders in two patents. This design had tightly fitting, fully closed, and intermeshing screws where the distance between flights

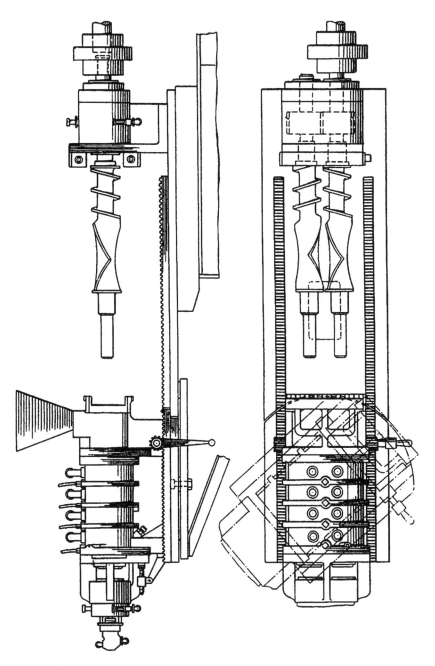

Figure 4 An illustration of nonintermeshing counterrotating twin-screw extruder from Ahlefeld. (From Refs. (15,16).)

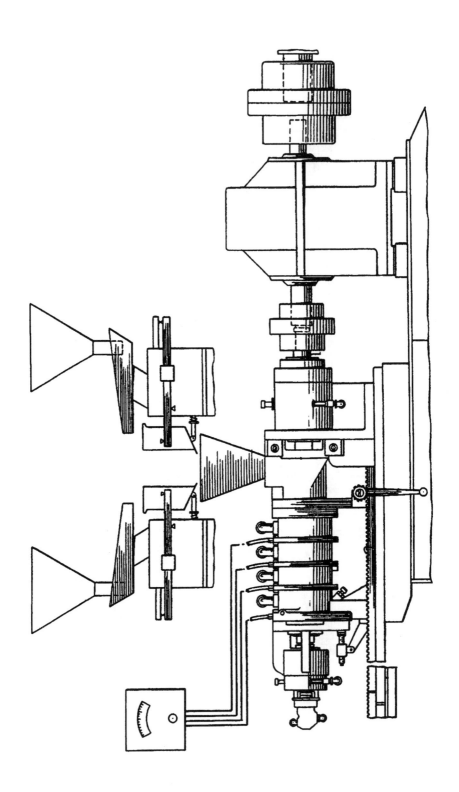

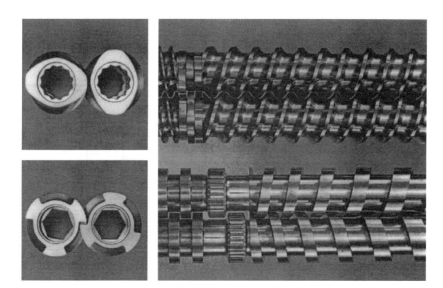

Figure 6 Twin-screw design examples: intermeshing corotating twin screw (top), and intermeshing counterrotating twin screw (bottom). (Courtesy of American Leistritz Co.)

steadily decreased in the pumping direction and resulted in a high shearing action (19,20). An illustration from the patent is shown in Figure 8. The first commercially available intermeshing counterrotating extruders came to the market in the 1940s.

C. Intermeshing Corotating Twin-Screw Extruders

Industrially, the most important type of twin-screw extruder is the intermeshing corotating twin-screw extruder. This type of extruder is also the most complex, due to the several different types of elements that are closely intermeshing and self-wiping. An illustration of an intermeshing corotating twin-screw is shown in the top example in Figure 6. Early origins of intermeshing corotating twin-screw extrusion are illustrated in patents in 1901 by Wunsche (21), and in 1917 and 1923 by Easton (22,23). An

---

Figure 5 An illustration of nonintermeshing counterrotating twin-screw extruder from Ahlefeld. (From Refs. (15,16).)

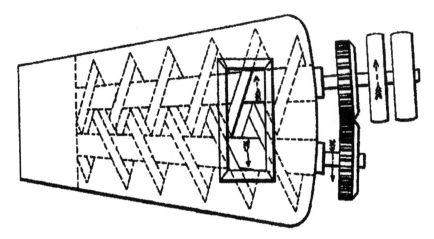

Figure 7 An illustration of intermeshing counterrotating twin-screw extruder from Wiegand. (From Ref. (18).)

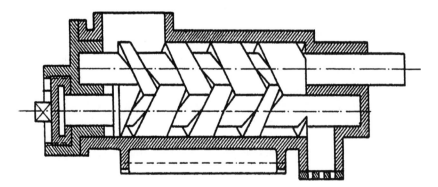

Figure 8 An illustration of intermeshing counterrotating twin-screw extruder from Leistritz and Burghauser. (From Refs. (19,20).)

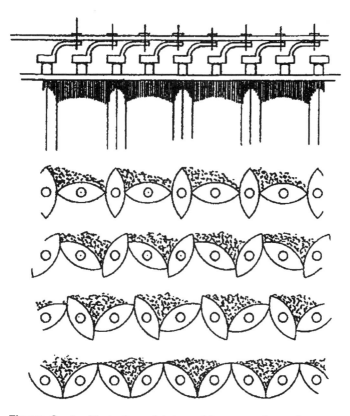

Figure 9 An illustration of intermeshing corotating twin-screw extruder from Wunsche. (From Ref. (21).)

illustration from the Wunsche patent is shown in Figure 9. The Easton patents states, "According to the invention… there are provided two intermeshing screws of similar pitch, which are of the same hand (either both right-handed or both left-handed) and which rotate in the same sense in close proximity to one another, and a casing which closely surrounds both conveyors". He later also mentions, "In all these applications of the invention, a great advantage arises from the fact that the surfaces of the two screws at the point of meshing are moving in opposite directions and thus they keep each other clean". An illustration from the patents is shown in Figure 10. Intermeshing corotating twin-screw extruders can achieve high speeds and outputs, while maintaining good mixing and conveying characteristics.

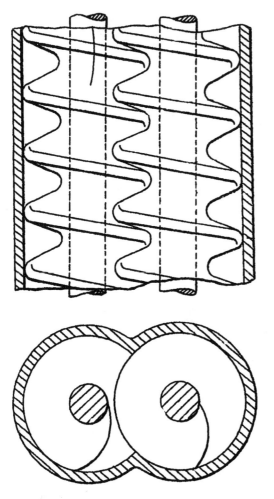

Figure 10 An illustration of intermeshing corotating twin-screw extruder from Easton. (From Ref. (23).)

VI. CONCLUSION

The processing with screw extruders has been used in many diverse industries for many years. The more recent introduction of twin-screw extruders with different screw designs allows them to be used to perform a large number of previously separate functions. The use of intermeshing screws permits a material to have a minimal residence time in the extruder, while achieving the desired mixing, melting, and conveying operations. Processing under the controlled conditions within an extruder allows temperature-sensitive products to be processed at temperatures higher than would be possible with conventional equipment. Applications of twin-screw extruders continue to expand in the pharmaceutical field, and the author anticipates that the same technical advantages that have led to widespread use in other industries will prove to have similar advantages in the pharmaceutical processing area.

REFERENCES

1. Bruin S, Van Zuilichem DJ, Stolp W. A review of fundamental and engineering aspects of extrusion of biopolymers in a single-screw extruder. J Food Process Eng 1978; 2:1–37.
2. Rossen JL, Miller RC. Food extrusion. In: Food Technology. 1973; 45–63.
3. Fielden KE, Newton JM. Extrusion and extruders. In: Swarbrick J, Boylan JC, ed. Encyclopedia of Pharmaceutical Technology. Vol. 5. New York: Marcel Dekker, Inc., 1992:395–442.
4. Janssen L. Twin screw extrusion. Chemical Engineering Monograph. Vol. 7. New York: Elsevier, 1978.
5. White JL. Twin Screw Extrusion: Technology and Principles. New York: Hanser Publ., 1991.
6. Reference deleted.
7. Sturges JD. Improvement in apparatus for cooling and mixing soap. US Patent 114,063, 1877.
8. Beck H. Injection moulding machine. German Patent 858,310, 1952.
9. Ferns AWD. Twin-screw machines for polymer compounding operations. Plastics Polym 1974; 42(8):149–157.
10. Rauwendaal CJ. Analysis and experimental evaluation of twin screw extruders. Polym Eng Sci 1984; 21(16):1092–1100.
11. Pfleiderer P. Innovations on kneading and mixing machines of Freyburger type. German Patent 18,797, 1882.
12. Reference deleted.
13. Thiele W, Petrozelli W, Lorenc D. Twin-screw compounding of slip

concentrates. A case comparison of intermeshing co-rotation and counter-rotation. SPE ANTEC Tech Papers 1990; 36:120–124.

14. Sakai T, Hashimoto N, Kobayoshi N. Experimental comparison between counter-rotation and co-rotation on the twin screw extrusion performance. SPE ANTEC Tech Papers 1987; 33:146–151.

15. Ahlefeld EH. Continuous internal stiff-gel mixer. US Patent 3,154,808, 1964.

16. Ahlefeld EH. Continuous internal stiff-gel mixer. US Patent 3,239,878, 1964.

17. Kaplan A, Tadmor Z. Theoretical model for non-intermeshing twin screw extruders. Polym Eng Sci 1974; 14(1):58–66.

18. Wiegand SL. Machines for sheeting dough. US Patent 155,602, 1879.

19. Leistritz P, Burghauser F. Kneading pump. German Patent 682,787, 1939.

20. Leistritz P, Burghauser F. Kneading pump. German Patent 699,757, 1940.

21. Wunsche A. German Patent 131,392, 1901.

22. Easton RW. Improvements in presses and pumps. British Patent 109,663, 1917.

23. Easton RW. Screw conveyer. US Patent 1,468,379, 1923.

2
Extruder Design

Richard Steiner
Leistritz Extrusionstechnik GmbH, Nuremberg, Germany

I. INTRODUCTION

Pharmaceutical class extruders are evolving as continuous processing devices to mix drugs with carriers for solid dosage forms and transdermal films, as well as to produce wet granulations. In melt extrusion, which has been used for many years in the plastics industry, the carrier is melted and mixed with an active ingredient, devolatilized and pumped through a die. In contrast, wet granulation refers to the process of fine powdered materials being mixed with liquids to impart morphological and other specific characteristics to facilitate the production and the performance of tablets and other solid dosage forms. The focus of the discussion will be primarily aimed at melt extrusion, the more common application and, to a lesser extent, at the wet granulation process.

II. GENERAL CHARACTERISTICS OF VARIOUS EXTRUDERS

Inside any extruder, a number of basic process functions are performed, which include feeding, melting, mixing, venting, and developing die and localized pressure. The motor facilitates the extrusion process via the rotation of the screw(s) that imparts shear and energy into the extrudate. Variable-speed a.c. drives are typically used. The gearing reduces the motor

19

speed to the desired screw rotations per minute (rpm) while multiplying torque. In the case of a twin-screw extruder, the distribution gear maintains the angular timing of the two screws, and absorbs the thrust load from the screw set.

Extruders process materials that are bounded by screw flights and barrel walls (1). Process control parameters include screw speed (rpm), feed rate, temperatures along the barrel and the die, and the vacuum level for devolatilization. Typical readouts include melt pressure, melt temperature, motor amperage, viscosity, and specific energy consumption (see Figure 1).

A common term used in extrusion is length-to-diameter ratio, or L/D. This is the length of the screw divided by the diameter. For instance, an extruder that is 1000 mm long with a 25-mm screw diameter has a 40:1 L/D. Typical extrusion process lengths are in the 24:1–40:1 L/D range. Single-screws are generally 36:1 L/D, or shorter. Intermeshing twin-screw extruders may be configured for up to a 60:1 L/D. The nonintermeshing twin-screw extruder can be specified at 100:1 L/D or longer, due to the absence of intermesh clearance constraints. Extruder residence times are generally between 10 sec and 10 min.

Other common terms include outside diameter of the screw, or OD. For instance, when referring to a 20-mm extruder, this refers to the outer diameter of the screw for a single-screw device, or the diameter of each screw for a twin-screw machine. The inside diameter, or ID, is the OD less the depth of the flight. The comparative OD/ID ratios determine the available free volume in any extruder. There are also various "gaps" in the extruder, either between the screw OD and the barrel wall (overflight gap), or, in the case of twin-screw extruders, between the screws (intermesh gap).

An important design factor in any extruder is the channel or flight depth. A deeper channel depth increases the free volume in the machine. It must be recognized that a deeper flight depth in the feed zone area decreases the screw shaft cross section and limits the possible torque transmittal. In the design of any extruder, it is important to find the optimum balance between free volume and torque, as this directly impacts attainable throughput rates, as well as the mass transfer energy that is imparted into the materials.

Barrels for extruders (single or twin) can be either one-piece or modular. The cross section of the barrel for a single-screw extruder reveals a circular hole for the screw, whereas the twin-screw extruder is characterized by a barrel opening in the shape of a "figure of 8". The inner surface of the barrel is tempered and honed. The barrel housing is made of high-strength steel, whereas the liner is manufactured from a wear-resistant and corrosion-resistant material as warranted by the intended service.

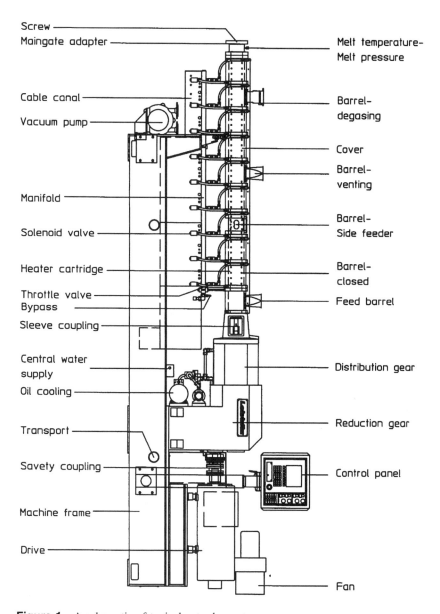

Figure 1 A schematic of typical extruder system.

The screw is generally deemed to be the most important part of any extruder. The screw design distinguishes the processes that the extruder can fulfill and, therefore, determines the quality of the extruded material. Similar to barrels, screws can be either one-piece or segmented. If segmented, the screws are assembled on shafts; usually either keywayed or splined. Splined shafts are considered "state of the art." Segmented screws allow for extreme process versatility, but may present cleaning issues in a good manufacturing practice (GMP) environment. One-piece screws can be selected to minimize cleaning issues associated with the disassembly of segmented screw elements from the shafts.

A fundamental condition for material viscosity and the chemical reactions that occur inside the process section are temperatures and mass transfer rates. The extruder barrel is typically heated, which can be accomplished by various methods. Conventional barrel heating is electrical, either by external heater bands or angular plates, or by internal electrical cartridge heaters. Barrels can also be heated by a liquid, such as treated water or oil, as dictated by the process and manufacturing environment.

Extruder barrel(s) is cooled via air or liquid, either externally or internally. The most effective heat transfer design uses axial cooling bores inside the barrel and close to the process melt stream. For certain applications, the number of cooling bores in the barrel may be increased to ensure better energy absorption. It is worth noting that twin-screw extruders dissipate up to two-thirds of the drive capacity into the product; therefore, it is essential to absorb this additional energy in a controlled way. Barrel cooling helps prevent material degradation and maintains the desired melt viscosity within the process section (see Figure 2).

Wear is caused by three mechanisms: corrosion, abrasion, and adhesion, often referred to as part of the "tribological system." The following is a brief description of each:

1. Corrosion is mainly based on reaction processes between the barrel/screw surfaces and the processed material, but can also result from high humidity and/or acidic atmospheres.
2. Abrasive wear occurs when screw metal rubs on barrel metal and/ or when the processed material between the screw and the barrel gap is abrasive. The process melt allows the screw(s) to "float" in the barrel to help prevent this type of wear from occurring. The more lubricating that material is, the more freely the screws may rotate and the lower the abrasive wear. Abrasive wear usually appears smooth.

3. Adhesive wear may occur when there is a loss of material presence between the screw and the barrel. Instead of abrading each other, interface pressures form tiny metallic fusion bridges, which then are almost immediately broken during screws rotation. For instance, this will occur if the identical stainless steels are specified with the same hardness. The result is a rough-looking circumferential surface. If these same materials are treated so that the hardness of the screw is slightly lower than the barrel, adhesive wear is prevented. Selecting different metals for screws and barrels is another common solution.

Figure 2   An end-view twin-screw barrel with internal cooling bores.

Specific energy is an important factor in troubleshooting any extrusion process because it indicates the amount of energy that is required to perform a process. Heater energy is usually not factored into this calculation. Specific energy is also an excellent troubleshooting indicator to determine if something in the process has changed. The formula for specific energy (SE) is as follows (4):

$$SE = \frac{Kwm \times EG\% \times TS\% \; (RPM_{run}/RPM_{max})}{Qh}$$

where

SE is the specific energy [kW hr/kg].
Kwm is the KW rating of the motor being used (KW=HP/1.3405).
EG% is the efficiency of gear system (0.954 is a reasonable estimate).
TS% is the torque used versus the maximum torque.
   TS%=amperes running/amperes maximum rating.
RPM_{run} is the running screw speed (rotations per minute).
RPN_{max} is the highest screw speed.
Qh is the output rate [kg/hr].

The specific energy of the extruder can readily indicate whether something has changed with the machinery or the formulation. Higher or lower specific energy might identify a material problem, or some electrical, mechanical, or processing parameter that has malfunctioned or has been modified.

III. VARIOUS EXTRUSION DEVICES AVAILABLE FOR PHARMACEUTICAL PROCESSING

There are various types of extruders that can be utilized to process pharmaceutical products. Each type has attributes and has been successful in demanding production applications for many years in markets such as medical device, electronics, packaging, construction, among others. Because other chapters are dedicated to the specific types, only a brief description is provided.

A. Single-Screw Extruder

The single-screw extruder is the simplest and most widely used extrusion system in the world. One screw rotates inside the barrel, which facilitates wide-ranging processes by means of various screw designs. The single-screw extruder is used for feeding, melting, devolatilizing, and pumping. Mixing is also accomplished for less demanding applications. Single-screw extruders can be either flood-fed or starve-fed, depending upon the intended manufacturing process.

B. Co-kneader

The co-kneader is a single-screw continuous compounding system that differs significantly as compared to a conventional single-screw extruder, due mainly to the axially oscillating movement of the screw. The screw is designed as a discontinuous flight with up to three equidistant interruptions or gaps per revolution. For example, in conveying sections, there is only a single interruption, while in mixing sections there are three gaps per revolution. Correspondingly, there are rows of fixed pins mounted along the barrel wall at equivalent spacing as on the screw flights. To avoid an interference between screws and pins, there is a complete forward and reverse stroke per revolution. The pins intermesh with the flights at the mixing elements so that the front and the back of each screw flight are wiped by the stationary pin. The kneading pins work like a static screw and are the counterpart to the rotating

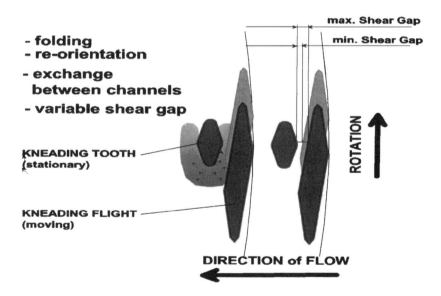

Figure 3 A co-kneader mixing mechanism (9).

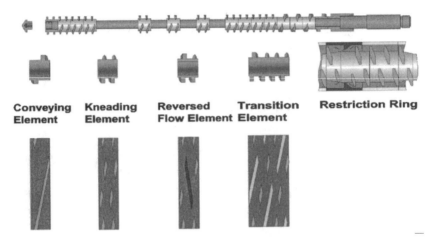

Figure 4   A co-kneader segmented screw system.

screw. Co-kneaders utilize a split-barrels design and segmented screws. These devices are typically starve-fed and use a discharge device, either screw pump or gear pump, to pressurize the die (see Figures 3–5).

C. High-Speed Twin-Screw Extruders

High-speed twin-screw extruders are mass transfer devices that are primarily used for compounding, devolatilization, and reactive extrusion, and are available in corotating and counterrotating. High-speed machines are defined

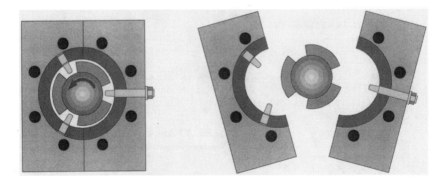

Figure 5   A co-kneader split-barrel design (9).

as those that have a top end of 300–1200 screw rpm capability, and are starve-fed with the output rate determined by the feeder(s). The screw rpm is independent from feed rate and is used to optimize compounding efficiencies. These devices typically utilize modular screws/ barrels, which offer extreme process flexibility.

High-speed twin-screw extruders are primarily specified in the corotating mode. One of the characteristics of corotating extruders is their ability to mix the material longitudinally as well as transversely. Consequently, the material is transported from one chamber of the screw to the other, which results in excellent mixing. Figure 6 shows the different streams of materials with various numbers of flights in the screw ($a=1$ flighted, $b=2$ flighted, $c=3$ flighted). The self-wiping of the two screws during rotation helps assure that intermeshing twin-screw extruders are self-cleaning. The intermesh of the screws also

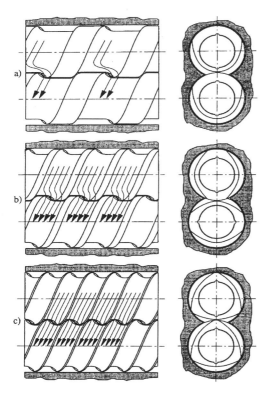

Figure 6   Corotating twin-screw flight designs (3).

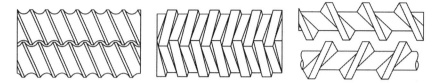

Figure 7 High-speed intermeshing and nonintermeshing twin-screw modes of operation—corotating/counterrotating.

supports the zoning of unit operations. Mixing primarily occurs in the kneading elements. Specialty high-speed counterrotating twin-screw extruders are available with up to six lobes for mixing, while tangential (nonintermeshing) designs are utilized for specialty applications (see Figure 7).

D. Low-Speed, Late-Fusion Twin-Screw Extruders

The low-speed, late fusion counterrotating intermeshing mode is characterized by a gentle melting and mixing effect that ensures a narrow residence time in conjunction with a high-pressure buildup, as compared to the high-speed twin-screw extruder that is specified for energy-intensive applications. The screw flights converge in the same direction in the nip region, causing the open spaces between the flights to be small, minimizing leakage from one screw channel to the next. This device can be used for applications where shear-sensitive or temperature-sensitive materials are being processed, or where high head pressures are desired, and/or where the materials do not convey well by

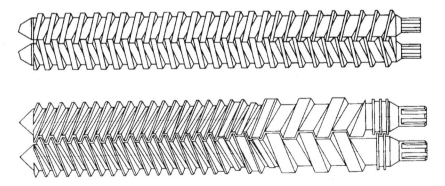

Figure 8 Late-fusion counterrotating screw sets.

drag flow. Corotating designs can be made to operate similarly through the use of nontraditional screw geometries (see Figure 8).

IV. WHAT IS THE DIFFERENCE BETWEEN A STANDARD AND A GMP EXTRUDER DESIGN?

The general process requirements and throughputs for the pharmaceutical industry are oftentimes less stringent than the demands of plastics processing.

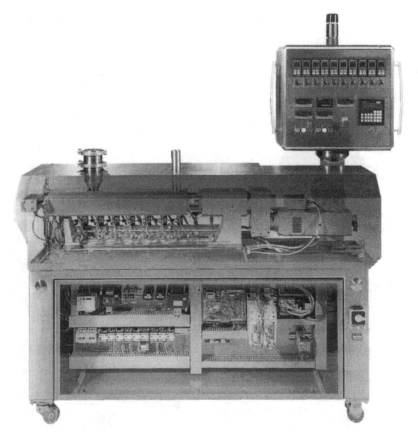

Figure 9 A GMP design twin-screw extruder design superimposed over standard machine.

A "pharmaceutical class" extruder is characterized not just by barrels/screws metallurgies, but also by special fittings and increased documentation, which are part of the "GMP philosophy." High degrees of sanitation, process reproducibility, and stringent documentation are typically important goals. Modifications, as compared to a plastics extruder, often include the following:

1. A polished, detailed design for all components with respect to cleaning
2. Machine frame and barrel cover in stainless steel
3. Ground and polished welding seams
4. Complete calibration and documentation of all process parameters
5. Quick-release couplings for cooling and heating system
6. The use of Food and Drug Administration (FDA)-approved gear oils and lubricants
7. Mercury-free pressure sensors
8. Materials in contact with products of stainless steel or nickel-based alloys
9. FDA-approved paints for parts that cannot be manufactured in stainless steel, such as the extruder gearbox
10. Validated programmable logic control (PLC) and computer-based controls.

Metallurgies for screws/barrels and other parts that contact the materials being processed are typically a hardened stainless steel or nickel-based steel, which are specified for FDA service so as not to be reactive, additive, or absorptive. The machine system should be configured for GMP cleanability and validation (see Figure 9).

V. CLEANING THE GMP EXTRUDER

There are two overall strategies for cleaning: the so-called clean-in-place (CIP) method, and a complete disassembly of the process section.

A. Clean-In-Place

Clean-in-place in the case of the extruder is not truly clean-in-place, because in all cases, the screw set must be removed from the extruder, but does not require a complete machine disassembly. Clean-in-place usually begins by purging the extruder with the screws turning at a low speed, typically 50 rpm or so. If the processed material softens or melts at an elevated temperature, the temperature setting in the process section should be above this temperature by several degrees.

The choice of purging material is dependent on the normally processed material, but many times a natural "solvent" can be found that will have a cleaning effect, and may be a liquid, such as a simple solution of water and detergent. The extruder is purged for several minutes with the discharge flange completely open, and the exiting material is collected in a container.

After purging, the extruder is stopped and the screw(s) is removed from the barrel. Normally the screw set is placed on a workbench for further manual cleaning. A set of two plastic "U" blocks made to fit the screws is helpful to keep the screw(s) from rolling off the workbench and also allows access underneath the screw(s). Depending on the residual material left on the surface of the screw(s), further cleaning may involve brushing with stainless steel wire brushes, wiping them with clean rags, and washing with more water/detergent. When the material residue is cleaned from the surfaces, the final step is to wipe down the screws with alcohol.

The first step in cleaning the barrels is to unbolt and to remove all the inserts for manual cleaning. A round wire brush attached to a long piece of stainless steel pipe is run up and down the barrel bore(s). If stubborn material deposits there, the brush pipe may be attached to an electrical drill and run up and down the bores while rotating. After this step, the bore(s) is cleaned further using cloth swabs. After the bores are free of all residues, the last step is to swab up and down the bores with a clean cloth soaked in alcohol.

If the machine is to be run again within a day or so, it can be left in this state for the next run. If it is to be stored for more than a few days, the screw and the barrel surfaces should be sprayed with an FDA-compliant anticorrosion coating. This is necessary because the intensive cleaning leaves the stainless steel surface vulnerable to corrosion, even in a humidity-controlled room.

Clean-in-place is designed to be able to prove that the machine is clean, but it is not as rigorous as a complete disassembly because there is the possibility of very small particles residing in the junctures. For this reason, most pharmaceutical companies have a policy of using CIP for perhaps three cleanings in between production runs, and at the fourth cleaning, a full disassembly of the process section is done.

B. Complete Disassembly/Cleaning of the Extruder Process Section

Before a full disassembly, the purging procedure as described above is still recommended to remove most of the material and to make the screw(s) easier to remove and clean. Prior to the teardown, the main electrical disconnect switch for the extruder should be turned off and locked out. A full disassembly involves unbolting all the barrel sections. For segmented screws, the elements are removed from the screw shafts. With a small laboratory-sized extruder, it may be easier to remove the entire barrel section from the machine as a unit, and disassemble the barrels on a bench. With a larger extruder, the barrel sections must be removed one at a time, starting at the discharge end. Once all the parts are on a workbench, all the surfaces are much easier to access.

Depending on the difficulty of cleaning the particular process material, the barrel(s) may be simply manually cleaned on the workbench with the heaters still attached. The heaters have to be removed if it is necessary to submerse the barrels in a bath of water/detergent, or some other liquid. Ultrasonic baths for barrels and screw elements have also been used to facilitate cleaning. As with the CIP method, the last step is to wipe all the surfaces with alcohol, and to apply a corrosion-inhibitor spray (see Figure 10).

Figure 10   A barrels removal device.

VI. SCALE-UP FOR PHARMACEUTICAL EXTRUSION SYSTEMS

Many applications involve scale-up from a few grams in the laboratory, mixed in small batches to a continuous extrusion process. The basic challenge is to properly choose those batch unit laboratory-scale operations that are best suited for use with the continuous extruder and its accessories.

It is extremely useful to have a laboratory extruder to benchmark very small batch subprocesses. From this starting point, dedicated processes can be configured for production at rates of 5, 10, 20, and 200+ kg/hr. It is difficult to generalize this translation due to the widely differing variables involved between products. However, a few rough guidelines follow (5):

1. Simple reactions are a product of time, temperature, and mass transfer to produce the same end result. This is experimentally, scientifically, and intuitively determined.
2. Distributive mixing requires the same number of same divisions per kilogram. This can be empirically calculated.
3. Dispersive mixing requires that the same product of stress rate and time per kilogram above the threshold level be created. This is calculated with some assumptions and empirical help.
4. Coarse devolatilization requires the same heat of vaporization with an adequate surface area. This is calculated.
5. Fine devolatilization requires the establishment of the proportional surface renewal. This is a combination of calculation and empirical data.

Factors that affect scale-up include volume, heat transfer, and mass transfer. Different processes are factored differently based upon which issue dominates. GMP extruder applications are generally more oriented to process precision rather than raw maximum output. However, execution and scale-up from batch processes and benchtop extruders remain governed by the same basic scale-up technology for other technical applications.

VII. PHARMACEUTICAL PROCESS EXAMPLES

Extrusion technology is an increasingly accepted method for the continuous processing of pharmaceutical materials, and often offers significant advantages as compared to batch processes. Upstream materials handling and downstream systems work in conjunction with the chosen extruder to perform the intended manufacturing operation. The following are a few examples of systems that have been successfully utilized in this capacity.

A. Wetting of a Powder Matrix with a Fluid

In this application, a powder is mixed with a liquid in a multifunctional feed port of a corotating extruder. The purpose of the fluid may be to lubricate, to become absorbed, or to facilitate a reaction. The feed port may be cooled to control the vaporization loss of the liquid. The powder, or another material, may also be added shortly downstream. After mixing is complete to the target morphological, rheological, or chemical condition, the material is discharged to downstream drying, compacting, or other equipment. Many of these materials cannot flow through dies. Discharge may be off protruding screws. Special features include the following:

> Early liquid injection and powder wetting
> Small pitch screws in the feeding section so that fluid does not move backward toward the gearbox
> Mostly conveying elements combined with some distributive mixing elements
> Screws protrude from the barrels to facilitate discharge to a downstream dryer.

B. Compounding of a Thermoplastic Urethathane Premix

A thermoplastic urethathane (TPU) premix is dried in a dessicant hopper/dryer and fed into a twin-screw extruder to extrude a product with antibacterial effects. Because atmospheric humidity causes the TPU to transform into a carcinogenic substance, an inert gas is introduced at the feed throat. Low-energy input pumping elements are specified for the second half of the process length to facilitate the removal of heat from the melt via barrels cooling for the heat-sensitive TPU. The mixing must be extremely homogeneous to optimize the antibacterial effect. Special features include the following:

> Inert gas atmosphere (N_2) for the feeder hopper and feed throat
> Extensive low-energy distributive mixing
> Intensive barrels cooling via internal bores to maintain a desirable melt temperature.

C. Devolatilization of a Cellulose Matrix

High levels of water vapor can be removed from a cellulose solution by means of a corotating extruder (7). Effective devolatilization is based upon the surface area of the melt pool, the surface renewal of the melt, and the residence time under the vent(s). Multiple stage degassing units along the extrusion process length are used. It is possible to remove 50%+ of volatiles via the extrusion process. Special features include the following:

> A screw design that incorporates multiple melts seals for vacuum and zero pressure under the vents
> Multiple vacuum pumps to facilitate volatile removal
> A gravimetric twin-screw feeder to meter moist cellulosic feedstock.

D. Incorporation of a Pharmaceutical Preparation into a Cellulose Matrix

The materials are mixed and devolatilized in a twin-screw extruder and mated to an air quench micropelletizing system (8). The in-line die minimizes stagnation in the machine front end and a self-adjusting cutter reduces the need for operator adjustment. Cold air nozzles can facilitate the cutting of materials that tend to smear at the die face. The microgranulate can be used to directly fill capsules without further size reduction. Special features include the following:

> A twin-screw extruder with extra process length to cool the melt prior to the die
> An air pelletizer with a cold air gun and a die assembly for micropelletization.

E. Incorporation of an Active Substance and Fluid into a Carrier Material

The materials are added in the following order: carrier material, fluid, and active substances. The active substance is fed into the extruder at a very late stage of the process to minimize the residence time in the machine to prevent thermal damage. Special features include the following:

> Multistage liquid injection with mass flow meters (see Figure 11)
> Air quench strand pelletizing system with a conveyor.

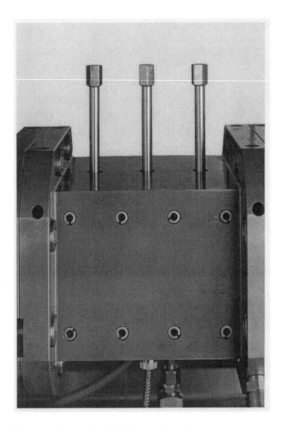

Figure 11 A liquid injection barrel.

F. Conveying and Compaction of a Tablet Premix

In contrast to melt extrusion, the screw geometry has minimal shear-inducing elements, and is designed with conveying elements with different pitches, which facilitates the conveying and the compaction of the powdered premix. Special features include the following:

A screw design that primarily conveys elements (except at discharge)

A special screw discharge to reduce stagnation before a die or a sanitary open release.

VIII. CONCLUSION

Single- and twin-screw extruders are replacing traditional batch processes because of the consistent and repeatable nature of continuous extrusion. Without addressing the specific application, it is impossible to state which type of extruder is best for the job. Fortunately, as evidenced by this text, there are many design options from which to choose. It is important to also recognize that the extruder is just another tool, albeit a powerful and versatile one, in the design of a continuous processing system for various drug delivery products.

REFERENCES

1. Ebeling F-W. Extrudieren von Kunststoffen. Würzburg, 1974.
2. Reference deleted.
3. Schuler W. Auslegung und Ausführung der Verfahrenszonen von gleichsinnig drehenden Zweischneckenextrudern. In: Der Doppelschnecken Extruder. VDI-K Verlag, 1995.
4. VDI K, Hrsg. Optimierung des Compoundierprozesses durch Rezeptur- und Verfahrensverständnis. Tagung Baden-Baden, 11. und 12.11.97, Düsseldorf, 1997.
5. Reimker M. Wärmezu- und Wärmeabfuhr. In: VDI, Hrsg. Grundlegende Aufbereitungsschritte von thermoplastischen Kunststoffen. Bericht vom Symposium am 21. und 22.4.98, München.
6. Reference deleted.
7. Thiele WC. Trends and Guidelines in Devolatilization and Reactive Extrusion. Chicago, IL: National Plastics Exposition, Society of the Plastics Industry, June 18, 1997.
8. Martin C. Continuous Mixing/Devolatilizing Via Twin Screw Extruders for Drug Delivery Systems. Interphex Conference, Philadelphia, PA, March 20, 2001.
9. Anderson P. Improved Design and Performance Characteristics of the Reciprocating Single Screw Extruder. Performance Compounding Conference, San Antonio, TX, April 10–12, 2002.

3

Single-Screw Extrusion and Screw Design

Keith Luker
Randcastle Extrusion Systems, Cedar Grove, New Jersey, USA

I. INTRODUCTION

Single-screw extruders are continuous, high-pressure-generating pumps for viscous materials that can generate thousands of pounds of pressure while melting and mixing. These devices are most often fed by pelletized solids with a regular geometry, such as spheroids and cylinders. Less common, and with somewhat greater difficulty, the feed may be diced material, powders, granulations, flakes, taffy, slurries, or melted material.

Mechanically, single-screw extruders are relatively simple machines. The screw is driven, ultimately, by a variable-speed motor, either a.c. or d.c. The motor drives a transmission that increases the torque and decreases the screw speed. Pressure, generated by the viscous melt being pumped through a die, commonly pushes against the tip of the screw. This exerts a backward force that is absorbed by thrust bearings.

The screw is surrounded by a barrel that resists high pressures. The barrel usually has three or more heating zones to raise the barrel and screw to the required process temperature. When the process temperature is driven above the temperature set point, barrel cooling, either air or liquid, cools the barrel.

In addition to the drive and temperature controls, there are usually various process readouts: pressure, stock (melt) temperature, and sometimes viscosity. Pressure is most commonly measured at the end of the barrel, and stock temperature and viscosity are measured between the extruder barrel and die.

Once the polymer is melted and mixed, it flows through a breaker plate with filter and shaping die, and thereafter is cooled into solid form.

Each process task, such as solids feeding, is associated with a particular section of the extruder screw, although several tasks are sometimes accomplished at once. A certain processing length, depending on the application, is therefore required to enable sufficient time for all the process tasks to be accomplished. Extruders are defined by the barrel diameter and length. Length is given as the L/D ratio or length divided by diameter. Typically, single-screw extruders are between 24/1 and 50/1 L/D.

The length given for each process operation depends on the tasks at hand. For example, a 24/1 general-purpose screw might have 4 equal-length L/Ds dedicated to solids transportation, melting, mixing, and pumping. A 30/1 L/D presents the opportunity to add 6 L/Ds to a critical part of the screw where additional feeding, pumping, melting, or mixing are required.

The output of the extruder is a function of screw diameter more than any other variable as output increases with the second power of diameter. Extruders range in size from 6 to 250 mm (and even larger). Because economy increases with the scale of production, most applications in the plastics industry use screws larger than 50-mm diameter. However, for pharmaceutical extrusion smaller extruders are used, typically with 6 to 30 mm.

Most extruder screws are driven from the hopper end. However, once screws are reduced to less than 25 mm, the screw becomes weak and solids transportation is far less reliable. To overcome these shortcomings, a vertical screw, driven from the discharge, may be used. The discharge of such screws is 2 to 4 times stronger, and the cramming action of the vertical screw increases solids transportation.

The extrusion of pharmaceutical materials is more challenging than the extrusion of typical plastics. Whereas any material, including drugs, can be extruded, pure drug formulations usually lack a key property required for uniform extrusion—the ability to flow. Without this, pharmaceutical materials tend to compact in the screw or have irregular discharge. Thus, the active ingredient is usually mixed with a component, often a biopolymer, to allow the mixture to flow along the screw.

Most conventional plastic feedstocks have additive packages mixed with the pure polymer to enhance processing. These additive packages have many purposes, such as reducing process temperatures, minimizing pressure variation, and limiting degradation. Because of the nature of the pharmaceutical industry and biopolymers, many standard additives cannot be used.

As a secondary consequence, biopolymers and pharmaceuticals may not be able to withstand multiple heat histories. This is important because, traditionally, many mixing applications take place in a twin-screw extruder where pellets are processed at low pressure (less than 1000 psi head pressure). This is then followed by single-screw extrusion where the product is extruded at a comparatively higher pressure (i.e., 3000 to 10,000 psi).

Many active ingredients cannot process at typical (150°C and above) extrusion temperatures. Therefore, carrier resins are selected to process within the drug's process range with regard to temperature and time. Such polymers are often lower molecular weight and may have a narrower weight distribution. Consequently, the extrusion process is more sensitive to small changes.

For those experienced with conventional polymer extrusion, pharmaceutical extrusion will seem slower, more sensitive to screw speed and temperature changes, and will take place with only one heat history. Nevertheless, once the process variables are matched to the materials, the process is extremely reliable and readily monitored to ensure consistent quality.

II. GRAVITY-INDUCED SOLIDS TRANSPORTATION

A. Hopper

In most single-screw extruders, solids transportation begins at the hopper where material is loaded. Ideally, the material is free flowing and will fall consistently from the hopper into the screw channel by means of gravity, known as "flood" feeding. Materials poured onto a flat surface and having an angle of repose less than 45° are often considered free flowing (Ref. 7, p. 169) and suitable for extrusion. Angles substantially greater have a tendency to compact within the hopper and may cause problems during the extrusion process.

Arching (Figure 1a) is the physical stacking of material into a natural arch as the material moves through the hopper or through the opening in the barrel above the screw. Some feedstock shapes, such as powders and granulations, are more prone to arching. The physical size of the opening to the screw is important because, given the same size feedstock (such as pellets) the larger the opening, the less likely arching will occur. So, the problem is more severe with the feed throat opening associated with small extruders. Hopper stirrers, preferably mechanical, can be used to break up the arches that form (Figure 1b).

If there are mixtures of physically dissimilar components, such as a biopolymer and a drug, the individual materials may separate because of the motion of the screw and the vibration of the extruder itself. Thus, one material

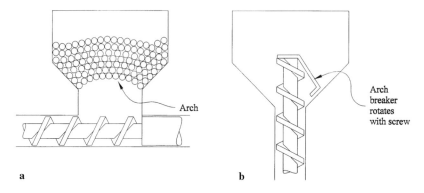

Figure 1 Arching (bridging).

(or the free-flowing component of a single material with a broad range of sizes) may flow preferentially into the screw, while the other component is accumulating in the hopper (Figure 2a and b). This will change the percentage of the components entering the screw and the mass flow into the extruder screw. Volumetric and gravimetric feeders can resolve the problem by feeding each component separately, as well as focusing careful attention to the preparation and quality of a premix.

After the material leaves the hopper, it enters the screw channel and solids transportation within the screw begins. Typically, the channel is of uniform depth in this part of the screw. Besides transporting material, the feedstock is usually preheated for preparation to enter the next section of the

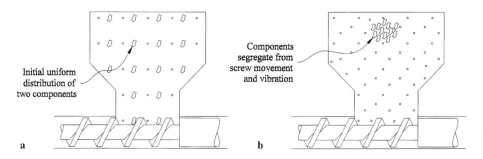

Figure 2 Component segregation over time.

screw. The feeding of solids is the least understood part of the extruder screw and can cause a great deal of difficulty.

A free-flowing feedstock composed of perfect spheres (typically 3 mm diameter) might be thought to produce exactly the same number of spheres in the first flight. The pellets, ideally stacking like an ordered pile of cannonballs filling the screw, would consistently take up much of the interstitial space between pellets (Figure 3a). The uniformly stacked pellets would then form a dense, regular feed that would continue down the screw. This would be highly advantageous because the objective is a continuous, stable process, and subsequent process tasks depend upon feeding.

In practice, this does not occur as feedstocks are not true spheres (varying on the order of ±20% in diameter and much more in mass) and, like ping-pong balls, do not behave as an ideal, free-flowing material. Instead, there is a chaotic filling where perfect uniformity cannot be expected from moment to moment. Nevertheless, if the extruder channel is at least 2 times larger than a pellet (Figure 3b), such as in a 50-mm extruder screw, and the differences in the spheroid geometries are relatively small, the filling will be sufficiently consistent for most requirements.

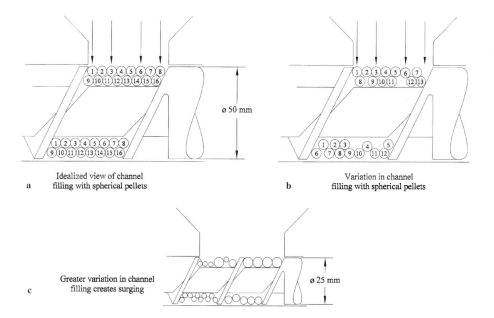

Figure 3 Mass variation: spheres.

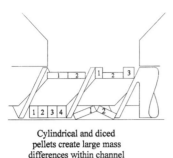

Cylindrical and diced
pellets create large mass
differences within channel

Figure 4 Mass variation: cylinders.

If a shallow channel is considered, as in a 25-mm screw, the mass can sometimes be fairly consistent (Figure 3c), but the pellets do not stack to remove much of the interstitial space. Given the typical difference in mass of the individual spheroids, relatively large differences in the mass of the first channel are created over time. Once such differences occur in the first part of the screw, the inconsistency is passed onward. This results in pressure and volume variations at the discharge of the extruder, commonly called "surging."

Regular-geometry feedstocks are available in cylinders, lens, and diced form. The bulk density of each form is different, which changes the mass flow characteristics. Although these feedstocks are all regular, dense geometries, the flow into an extruder screw channel is not as consistent as spheroids. Instead, these stack momentarily and then collapse with the motion of the screw. These differences are exacerbated in smaller screws (Figure 4).

Some feedstocks are irregular (irregular when magnified) such as powders and granulations. For a variety of reasons,* these materials are very popular in pharmaceutical extrusion. As mentioned, these materials are prone to flow differences from the hopper. In addition, these materials sometimes exhibit arching in the channel itself as the particles interlock. If an arch forms, the mass flow into the channel will be prevented as it blocks material from entering the screw from the hopper (Figure 5). Besides interlocked arches, trapped air pockets and static electrical charges can suspend material and result

---

* Reasons include the desire for a single heat history, granulation of standard pellets so that drugs can be distributed in the feedstock, better mixing of higher aspect ratio materials, and availability of reactor-grade materials.

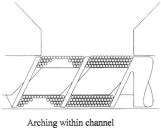

Arching within channel
creates large mass differences

Figure 5 Channel arching.

in filling problems. Materials with low bulk density, less than 2 g/cm^3, will also likely cause filling problems (Ref. 7, p. 168).

The problems associated with uniform filling of the channel explain the popularity of pellets as the preferred geometry. The most important features of successful powders and granulations are free-flowing character into the screw, sufficient friction to transport once inside the channel, and narrow particle-size distribution.

B. Solids Conveying

It is essential, at this point, to understand the mechanism of solids transportation. In Figure 6, the barrel revolves around a stationary screw (although in practice the screw turns). When the pellets enter the channel, the barrel must

Figure 6 Friction and conveyance.

contact the pellets and drag them against the flight so that transportation takes place. Thus, it is friction between the barrel and pellets that is required to overcome the drag on the screw.

Another way to conceptualize what occurs is to visualize a nut on a screw (the nut represents the pharmaceutical material) and that the screw is slightly rusted, making it difficult to hold the nut with a wrench. If the nut slips from the restraining force (the wrench), the nut then revolves *with the screw* and the nut will no longer move forward. Anything that makes the nut slip from the wrench (i.e., oil or wear on the wrench) will retard the nut's advance. Likewise, anything that makes the barrel slippery (reducing the friction between the material and the barrel) will retard the advance of the process materials along the screw.

If friction is slightly reduced, e.g., by the addition of a tiny amount of fluid (such as oil, water, or plasticizer), the ability of the barrel to drag material forward is reduced. As the amount of lubricating agent is increased, barrel drag decreases until the material sticks to the screw completely and forward transportation stops.

Besides lubricating agents, there are other instances where barrel friction becomes a problem. If the feedstock's density is too low in the channel (such as in the case of ground film or fine powders), then the barrel will not have sufficient friction to drive material forward. If the material itself is too slippery, then material will not advance. If the feedstock is melted prematurely—in what should be the solids conveying zone—then the lowered friction may not drive material forward along the screw.

C. Gravity-Fed Solids Feed Section

The feed section of the barrel is a distinct, exchangeable section that is typically water cooled. A separate feed section allows for thermal insulation between barrel and feed section and prevents heat from prematurely melting material in the solids conveying zone and stopping transportation. The feed section should have a cooling length of 3 L/Ds between the heated barrel and the feed throat opening (Figure 7a, b, and c). Furthermore, control of the feed-section temperature facilitates changes in friction to the feedstock. Thus, temperature control permits small adjustment to mass transfer and minimizes surging.

The most common feed section used with gravity filling of horizontal extruders (Figure 8a) is a round or rectangular opening over the screw

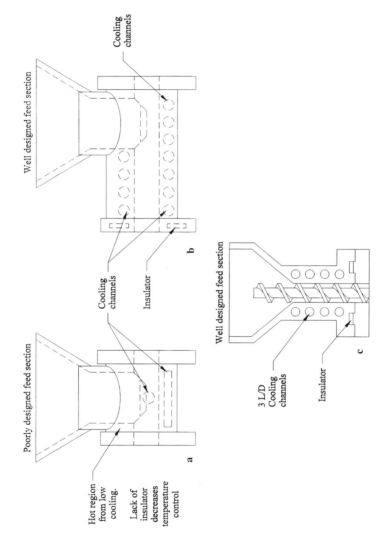

Figure 7 Feed section cooling length.

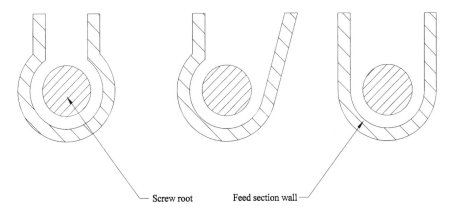

Screw root   Feed section wall

Figure 8   Horizontal feed section geometry.

centerline. This type of feed section is entirely adequate for standard-size pellets in extruders greater than 25-mm diameter. However, with pharmaceutical extrusion the standard design, merely scaled down, often leads to feed problems, as will be discussed later.

Sometimes, in an attempt to overcome arching in small extruders, the feed opening is enlarged (Figure 8c). However, this reduces friction (barrel holes do not convey and larger holes convey more poorly) and solids transportation.

Temperature control of the 3 *L/D* feed section (Figure 7b and c) allows the friction in this barrel section to be altered either by the amount or temperature of the coolant (usually water). Because friction changes with temperature, this region is often critical for consistent solids conveying. Thus, good control of the feed section temperature drives the correct amount of material forward for proper melting and pumping.

A shortened feed section reduces the opportunity to control friction and makes it more difficult, sometimes impossible, to transport material. Figure 7a shows a poorly designed feed section without 3 *L/D* cooling downstream of the inlet. This type of feed section will be cold on the bottom and hot on top, making consistent feed treacherous.

Below 25 mm diameter, horizontal extruders can sometimes be problematic. First, variations in the size of individual pellets (or other feedstock) change the mass transport over time. Second, chaotic variations in how the pellets fall into the channel results in different numbers of pellets even when the pellets are about the same size.

D. Gravity-Fed Melt Feed Section

Less commonly, molten materials are fed in tangential feed sections (Figure 8b). These should not be used for solid feedstocks because of the lateral displacement of the screw from the wedging action of the feed.

E. Pressure-Induced Solids Transportation

Often, gravity is not sufficient to fill the channel and Archimedean transport results in a partially filled channel (Ref. 7, p. 234) (Figure 9a). If this partial filling is consistent and the extruder channel is large, then the mass transfer will still be reasonably consistent. However, this type of solids conveying can exacerbate the mass flow differences in a small channel (Figure 9b). In such a case, the difference in mass is 6.5 times greater. Such differences in the initial filling of the flight will contribute to surging at the discharge.

Consider the consequences of "pushing" on the material being fed instead of relying on gravity. The channel filling will change because the

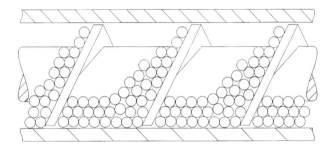

a   Archimedean flow in large screw gives reasonably consistent flow

Low mass   6.5 times greater mass

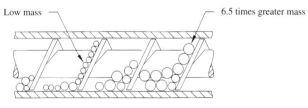

b   Archimedean flow in smaller screws creates mass flow differences and surging.

Figure 9   Effect of screw size on Archimedean flow.

material will be packed into the screw. Several techniques are possible including crammer feeders and low-pressure feed sections.

1. Crammer Feeder

Feedstocks may be densified before reaching the screw by "pushing" the materials into the channel by a crammer feeder (Figure 10). This is a screw in the hopper that compresses material to deliver a denser feed to the first screw channel. The densified feed also raises the pressure slightly in the channel, producing a slight load on the solids conveying zone. Crammer screws are most commonly employed in the blown-film industry where shredded plastic bags are reclaimed. However, crammers are potentially useful in pharmaceuticals where the density of the material may be too low to feed without assistance.

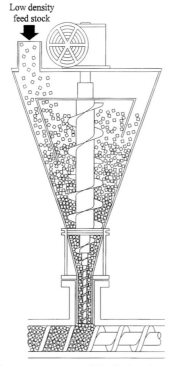

Crammer increases mass filling in channel

Figure 10 Crammer feeder.

Note that in a crammer feeder, friction is still the driving force on the material. Therefore, the most lubricious and lowest density materials may not feed better in a crammer feeder than with a regular screw.

2. Roll Feeders

Some feedstocks, such as silicone and "taffy"-like materials, are pushed into the screw with a rotating roll (Figure 11). The rotating roll forms a nip point between the roll and outer surface of the flights. Typically, this barrel section is undercut so that material is forced into the screw channel. This type of enhanced feeding also produces a slight pressure in the solids conveying zone with an associated load. These feed sections are most commonly used in the rubber industry to feed strips of material.

3. Low-Pressure Barrel Feed Sections

In a vertical screw, the continual advance of the flight continually pushes material into the flight at low pressure. This packs even differently sized pellets together (Figure 12) and creates a more uniform mass flow than Archimedean flow.

Because the feed section of the barrel is easily changed in the vertical design machine, several, smooth-bore feed sections are available. It should be

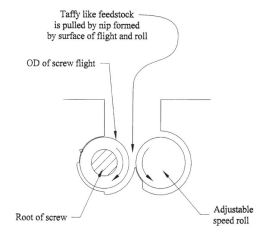

Figure 11   Roll feeder.

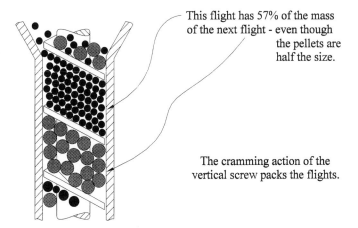

This flight has 57% of the mass of the next flight - even though the pellets are half the size.

The cramming action of the vertical screw packs the flights.

Figure 12   Low pressure feeding in vertical screw.

noted that small extruder screws feed very few pellets per revolution. A 16-mm screw, for example, might feed 10 pellets per revolution. Thus, feed sections that transport only one or two pellets more per revolution become useful because it represents a 10% to 20% increase and prevents surging. It has also been shown that pressure stability is influenced by pellet material, shape, durometer, and other factors (3). Because solids transportation is poorly understood, the optimal feed section is often found by experiment. This is easily accomplished in the vertical design because the feed section is changed without die or screw removal.

4. High-Pressure Barrel Feed Sections

In the feed sections discussed so far, the bore of the feed section is essentially smooth. There is a significant amount of slippage as the pellets tumble and slide along the screw. The pellets act like ball bearings where the contact of the barrel to the pellets' surface is small.

A high-pressure barrel feed section can be created by adding one or more grooves to the barrel bore under the hopper and into the solids conveying zone. The grooves are usually parallel to the screw axis (Figure 13a and b). Typically, the grooves extend about 3 *L/D*s beyond the opening, although high pressures can be generated with as little as 1 *L/D*.

In gravity and low-pressure feed sections (assuming a channel is suffi-ciently filled), material advances along the screw flight and flows in a spiraling

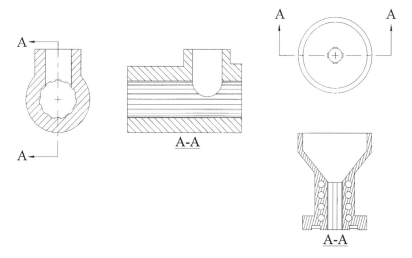

Figure 13 Horizontal and vertical grooved feed sections.

fashion. However, in a straight-grooved feed section, the pellets are trapped in the grooves and advance within the grooves as the flight pushes on the pellets. The result is greatly enhanced transportation along the screw axis. In turn, the pressure in the early part of the screw is increased. The pressure compresses the feedstock into a solid bed within a short distance as temperatures quickly increase. The screw design must reflect the increased transportation by increasing the downstream channel volume, as compared to a standard screw.

It should be noted that grooved feed sections, in the majority of applications, only feed pellets. Most powders will fill the grooves making the feed section function as a smooth feed section or with just a small increase in throughput.

The main advantages of grooved feed sections are the following: solids conveying will result in very high pressures to overcome high die resistance, uniform pressures at the screw discharge, the ability to process slippery materials, and higher output per revolution for a given size screw. Disadvantages include the following: reduced mixing as the active/carrier may be compacted prior to melting; generally, the inability to process forms other than pellets; and high wear in the grooves.

For pharmaceutical extrusions, the disadvantages of grooves may outweigh the advantages when mixing is paramount and where pelletized feedstocks are not available.

III. MELTING/PUMPING/MIXING/VENTING

There are two schools of thought about melting and mixing. Historically, single-screw extruders have relied on the theory that a screw first transports solids; that the solids are preheated; then melting takes place in the decreasing space of the compression section of the screw (Figure 14); then the melt is pumped through a filled/pressurized mixing element on the screw.

Therefore, materials are pushed through a mixer at relatively low viscosity (compared to the viscosity as material was beginning to melt) and *sheared* within the mixer to disperse the components. In this view, single-screw extruders are pumps with little mixing, as compared to other devices, such as twin-screw extruders.

An easily understood pamphlet is published by Spirex (1) that shows many of the various possible screw designs. A mathematical distillation of common barrier screw designs is given by Rauwendaal (Ref. 7, pp. 403–415). However, there is little comparative published data based on actual experiments.

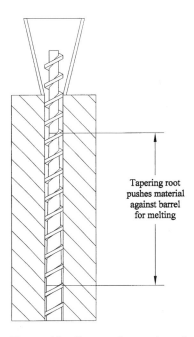

Tapering root
pushes material
against barrel
for melting

Figure 14   Compression section of screw.

This is not the only possible way to design a mixer or operate a single-screw extruder. There is evidence that single-screw mixing might be significantly better if it employed elongation to at least partially elongate material as it moves through a mixer (4,9) (Ref. 8, pp. 219–227). Further, because it is well known that mixing improves at higher viscosities, such as when melting first begins, elements can be designed to melt and mix simultaneously. In this view, melting and mixing can occur at the same time, as will be seen shortly.

A. Melting

Once the feedstock has been transported through the solids conveying zone and is preheated, there is enough energy available for melting. The decreasing channel volume compresses the material as it moves through the decreasing space of the melting zone. This forces the air between the feedstock particles

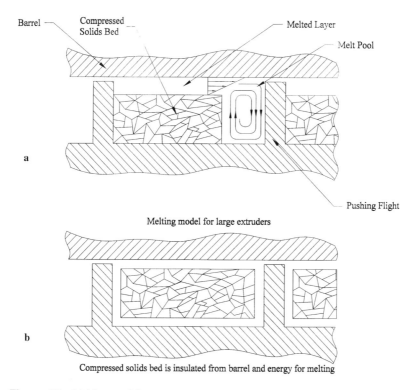

Melting model for large extruders

Compressed solids bed is insulated from barrel and energy for melting

Figure 15 Melting models.

out through the solids conveying region and forces the material against the barrel where melting occurs.

Melting in this mode of operation takes place as the metal barrel conducts energy into the polymer, forming a melted layer (Figure 15a). The advancing flight then scrapes the polymer into a rotating pool on the leading edge of the flight. In larger extruders (over 60 mm), the channel can become so large that the solid bed can become completely isolated from the barrel, allowing unmelted material to leave the screw (Figure 15b).

The most common solution to the problem is a melt separation screw. There are a variety of designs available, but the basic principle is shown in the Uniroyal design (Figure 16). Here a secondary flight is introduced with a greater clearance to the barrel as compared with the main flight. The secondary flight is referred to as a "barrier" flight. Because the solids pressure is higher than the melt pressure, the melted material is driven over the barrier. The melt is thus drained away from the solids bed, allowing the solids bed to contact the barrel and continue the melting process.

A common problem associated with this design is compaction ahead of the barrier flight. When this occurs, the solid bed "jams" at the entrance to the barrier. This can lead to loss of output, material stagnation, and degradation. Research by Dow Chemical (2) demonstrated that removal of a portion of the first part of the barrier flight could effectively solve the problem. This modification often results in reduced process melt temperatures and increased output without degradation. In addition, mixing should improve because active ingredients tend to not be compacted prior to mixing.

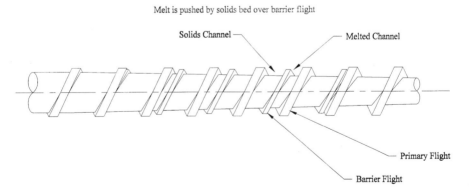

Figure 16   Barrier screw.

B. Pumping

Because pressure variation at the end of the melting zone is too great for most applications, a metering section is used to dampen out pressure fluctuations. The section is also called the pumping zone and is usually a uniform channel height. Once the melted material fills the metering channel, then the metering section acts to build pressure to overcome the die resistance.

It can be seen from the following equation (10) that pressure is a function of channel depth and length.

$$\Delta P_{max} = \frac{6\pi \cos \theta_b}{\sin \theta_a} \frac{ND_b\mu l}{H^2}$$

where:

P	=	pressure
θ	=	helix angle, rad
θ_a	=	average helix angle, rad
N	=	frequency of screw rotation, rps
D	=	diameter, cm
μ	=	viscosity, dyne sec/cm^2
l	=	axial length/cm
H	=	channel depth=distance between screw root and barrel surface, cm

The shallower the depth, the higher the pressure generation to the second power. Longer metering sections produce proportionally higher pressures, everything being equal.

The flow through the metering zone is not axial but forms a spiral within the screw channel (Figure 17). Because of the spiraling motion, not all the

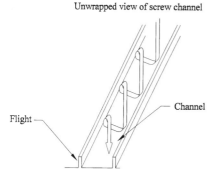

Figure 17 Metering section flow.

material moves through the extruder screw at the same velocity along the screw axis. The material in the middle of the spiral will move more quickly through the screw. This accounts for some of the residence time distribution of the single-screw extruder and the axial mixing along the screw.

When the instability of the melting zone cannot be sufficiently damp-ened by the metering section and process adjustments, a gear pump front-end attachment can be added at the end of the extruder. This can reduce pressure variations to around 10 to 20 psi and volumetric fluctuation to less than 1%. Pressure variation can be similarly reduced by automatic screw-speed control in the presence of a surge-suppressing seal (5,6).

C.   Vented Extruders

Once the material is melted, it is sometimes pumped into a decompression zone and processed through a second screw section to allow the extraction of gases such as water, residual monomer, and other volatiles from the extruder. The screw sections are designed so that the output of the first stage is less than the second section (Figure 18). In this way, the input of the second screw will be at zero pressure, which allows the extraction of gases, often under a vacuum, to enhance gas removal through the barrel. The pumping balance between sections can be delicate and vent flooding (excess material coming out the vent instead of gases) can occur. The length required for this process task typically adds 12 L/Ds so that a 36:1 L/D is about the shortest length for a well-designed two-stage screw.

Tandem screws solve the flooding problem because the balance between screws can be adjusted. Further, the very long overall L/Ds mean better venting. However, this is at the expense of a second drive, screw, and barrel.

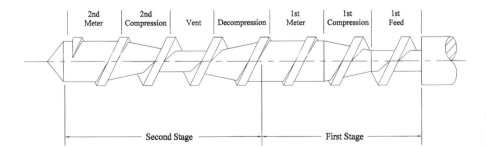

Figure 18   Two stage screw for venting.

D. Mixing

Mixing increases the tendency toward uniformity in a formulation. Mixing, therefore, refers at least to temperature, liquid, and solids uniformity. Generally, there are three types of mixing commonly referred to: dispersive (mixing by breakdown of solids above a yield point), distributive (mixing by reorganization of components toward uniformity and below yield point), and thermal homogenization (distributive mixing in liquids).

When dispersive mixing is present, distributive mixing also naturally occurs. The reverse is not true.

Single-screw mixing elements are not readily quantified. As a result, one practice is to qualify mixers based upon generalized mixing principles. This suffers from the many assumptions of the reviewer. Another practice is to quantify the results of experiments using a particular mixer. Although this produces concrete results, there are no industry standards and so comparisons are difficult. Pharmaceutical extrusion mixing studies are virtually nonexistent. In reviewing the screws available to mix pharmaceuticals, we are forced to rely on general engineering and polymer experience.

As noted earlier, historically, it has been most common in single-screw extrusion to add the mixers in the latter part of the extruder screw. This is logical because most mixers do not generate pressure, but consume pressure and, therefore, have been placed after the pumping zone. This places the mixer after the material is fully melted and the viscosity, therefore, near its minimum. This placement can be especially useful for melt homogenization to reduce the thermal gradients in the melt as it leaves the screw.

Mixing may also be enhanced by the type of feeding. Most single-screw extruders are flood fed. That is, the hopper sits over the feed throat and gravity fills the screw channel to its limit. Solids are then compressed and melted to fill the end of the screw.

However, compression of solids is not necessarily consistent with optimized mixing. Consider a compressive extruder screw where equal amounts of a pellet, "A," and a pharmaceutical material, "B," are to be mixed in the extruder. On entering the screw, it is extremely unlikely that the mix will be perfect, i.e., A-B-A-B-A-B-A-B etc. Some inconsistent sequence will likely occur such as A-A-A-A-B-B-B-B. High compression of this sequence at high pressure often results in very sturdy "A" and "B" where each agglomeration has to be broken up, i.e., mixed by dispersion, and then mixed intimately by distribution. This is inherently counterproductive.

Generally, the higher the compression or the sooner compression takes place along the length of the screw, the worse the problem created by compression will be. Grooved feed extruders, because they create high pressures very quickly, generally create the worst problem, followed by low-pressure feeding, flood feeding, and starve-fed extruders.

During starve feeding, the hopper is not filled. Instead, another device limits the amount of material that falls into the screw to less than would occur during flood feeding. The feed limits are adjusted by volumetric or gravimetric feeders. As material falls into the screw, the feedstock is pushed against the leading edge of the screw providing forward transportation. Because the screw is not filled, pressure is very low, and remains so as the material is preheated. The preheated material is then delivered to the mixer with as little pressure as possible.

Starve feeding is generally helpful in another respect. Because starve feeding moves material through a mixer at a reduced rate (compared to flood feeding), the mixing action takes place over a longer period of time. Whereas this will lower the output per revolution, the screw speed and L/D ratio can be increased to compensate, if warranted.

Whereas single-screw mixers have been designed to use shear flow, elongational flow can also be utilized. Elongational flow has been shown by White (12) and others to be more effective for dispersion. Thus, mixers that employ substantial elongational flow can be effective under certain conditions.

The advantages of elongational flow for distributive mixing are easily understood. When an imperfect sequence of polymer and pharmaceutical enters the extruder (i.e., A-A-A-A-B-B-B-B), during extensional flow the components are moved farther apart with the opportunity to drive the mixture toward uniformity rather than toward agglomeration.

Dispersive mixing requires a high stress rate—the product of the controlling modulus (the viscosity) times the strain rate (11). The highest viscosity is when the material is nearest its solid phase. It should be noted that the viscosity of most materials at this condition is orders of magnitude higher than during the melt phase. Therefore, the greatest dispersive forces can be generated when mixers are placed nearest the solids transportation zone where the material is in a semimolten form.

The following describes mixers used in single-screw extruders:

Mixing Pins.   A common mixer placed at the end of the screw is a series of pins that act as obstructions to the spiral flow of the metering section and where the flow is split and recombined. Pins of just about every imaginable shape are available including round, square, pyramids, trapezoids, and diamond-shaped pins. The pins may be placed within the flight channel in the metering section or after the flights are stopped (Figure 19).

Figure 19 Mixing pins.

Because of low-flow regions behind some pins, such as round and square, degradation can occur. Small round pins can break and shear off because they weaken as they flex during surges and cold start-ups. Round pin mixers should be avoided for pharmaceutical extrusion because of degradation in the low-flow region behind the pins.

Slotted Flight Mixers. Another class of distributive mixers is slotted flights where one or more flights are made discontinuously. These are historically best represented by the Dulmadge (Figure 20a) and Saxton mixers (Figure 20b). In terms of pharmaceuticals, the Saxton might be slightly better because it has some pumping capability and barrel wiping.

A relatively new slotted mixer is available that uses elongation of the melted material to achieve mixing. Rauwendaal has made a mathematical analysis and filed a patent on the CRD mixer (Ref. 8, pp. 221–224). The analysis of the material passing through the curved flight flank indicates that it is subject to elongational flow and, therefore, dispersive mixing. Slots between the curved flanks are created to increase distribution, and also expose material to elongational mixing. While the slots will increase distribution, they will

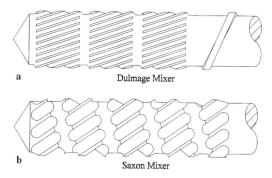

Figure 20 Slotted flight mixers.

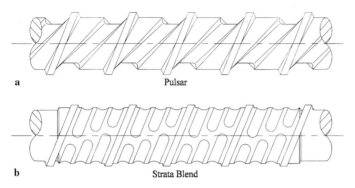

a   Pulsar

b   Strata Blend

Figure 21   Variable channel depth mixers.

reduce the amount of material passing through the elongational action—the primary feature of the mixer. Therefore, the mixer is now made relatively long—about 6 L/Ds—in order to ensure that most of the material may pass through the elongational action.

Because the slotted CRD is relatively new, little is known about the mixer except for the mathematical analysis. In any event, because the mixers have been placed at the end of the screw, the controlling modulus will still limit its effectiveness for dispersive mixing.

Mixers That Shift Material Within Flights.   Shifting material within the channel is accomplished either by variations in the channel height or by variations in a barrier within the flight. Commercial examples include the Pulsar by Spirex (Figure 21a) and the Strata-blend by Newcastle (Figure 21).

This group is less likely to create as many flow streams as pin mixers or slotted flight mixers. However, the shifting of material toward the barrel is potentially advantageous for melt homogenization. Further, these mixers are gentle and might be useful for sensitive polymers and drugs that cannot withstand high shear.

Fluted Mixers.   Fluted mixers are generally believed to be more dispersive than distributive. The most common slotted mixers are the Egan (Figure 22a) and Union Carbide (UC) mixers (Figure 22b). Material is forced down an inlet channel that dead-ends at a radial barrier where shear takes place. Thereafter, the material is forced through an outlet channel. The Egan mixer is angled so that it is not a purely pressure driven device, as is the UC mixer.

These mixers are very popular in the plastics industry but are limited to shear mixing in the small radial clearance. This mixer does not create many

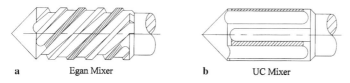

Figure 22   Fluted mixers employing shear.

recombining flows. These mixers do have some melt homogenization benefits as the material rolls through the inlet and outlet grooves where it is exposed to the barrel surface and energy is exchanged.

In terms of pharmaceuticals, the inlet and outlet channels should not be made with an abrupt end to the inlet channel or the beginning of the outlet channel as this can contribute to stagnation. Instead, a tapered channel is preferred. However, this will also decrease distributive mixing.

Another fluted mixer, the Randcastle BT mixer (Figure 23), is similar to a UC mixer but is distinguished by an open-ended inlet channel and by its location. For distributive mixing, it is located near the end of the screw but should be followed by a pumping zone.

In the BT mixer, the radial shear plane becomes a pump. For the pump to be effective, the inlet channel must be starved below the pumping rate of the radial clearance (the pumping land or cross-axial pump). As material is pumped axially (by pressure flow) and is pumped cross-axially (by the

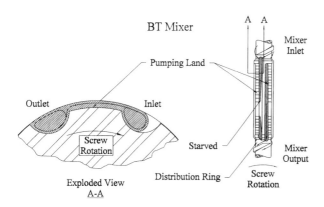

Figure 23   Recirculating fluted mixer employing elongation.

pumping land), the mass flow is elongated promoting mixing and melting. Because the viscosity is near its minimum, the L/D of these distributive mixers is usually 4 to 5 L/Ds.

Once material has been pumped through the outlet channel, the material discharges to an area where it forms a ring. Because of the rotation of the screw, the ring material will be pumped circularly and this becomes the source of material that reenters the mixing inlet channel. This material is again pumped down the inlet channel, opposite the main flow direction, and again experiences elongation.

This material, having now received a second mixing effect, combines with material coming down the output channel having only one mixing history. The material leaving the outlet channel, part having one mixing history and part having two mixing histories, will then be pumped into the distribution ring. The process will then repeat (Figure 24).

This process will not form an infinite number of repetitions but will reach equilibrium. The number of repetitions is primarily determined by the size of the ring (the bigger the ring the more repetitions) and the degree of starve feeding (the less feeding, the greater the number of repetitions). When more material flows into the ring than flows into the mixer, material will

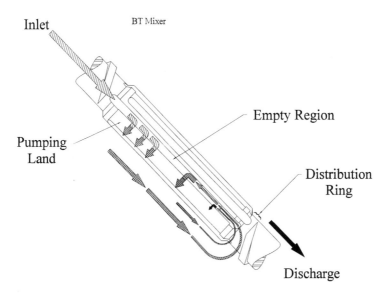

Figure 24 Recirculating fluted mixer employing elongation: detailed view.

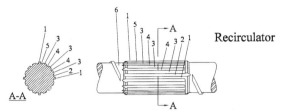

Recirculator

A-A

Triple Pass Recirculator: This Recirculator triples the mixing action
of the BT Mixer. The numbers above show one quadrant of the mixer.
Material is divided into four streams entering channel (2) bounded by the
wiping land (1). Material is subject to strong elongational flow as it is
drawn away by pumping land (3) and pumped into channel (4) where the
mixing is twice repeated until exiting at channel (5). Thereafter, material
re-enters channels (4) and (2) at the distribution ring (6) and re-mixes.

Figure 25 Recirculating fluted mixer employing multiple elongational flow paths.

reach the flight and will advance down the screw. Because this mixer creates
a very large number of layers, it is good for melt homogenization.

A second patent-pending version is called the Recirculator (Figure 25),
by Randcastle. In this variation, the mixer has more than one channel after the
first outlet channel where material can again be elongated. This has the
advantage of creating secondary elongational flows within the mixer and
before the controlling modulus can drop significantly. The secondary mixing
permits additional layer creation over a shorter axial distance.

Barrier Mixers. There are two similar screws that use barriers to melt.
These are the Double Wave screw by HPM (Figure 26a) and the Barr
ET®screw (Robert Barr, Inc) (Figure 26b).

In the Double Wave screw, the channel depth is reduced on one side of
the barrier and increased on the other. Thus, material is pumped over
successive barriers and sheared. This occurs twice per L/D. The Double Wave
screw is often utilized with a long, straight preheating section before the mixer
to prevent compression.

The Barr screw is similar, but the barrier clearance is much larger at
about 3 mm. The smallest screws using this design are in the 50-mm-diameter
range. The design is said to offer lower melt temperatures by mixing the
temperature gradients at low shear over the barriers (distributive melt mixing).

As previously noted, the higher the viscosity, the higher the stress rate
and dispersive mixing. Because the viscosity is many times higher when
melting first begins, dispersion can be significantly improved when mixers are
placed early in the screw. This will combine mixing and melting and has

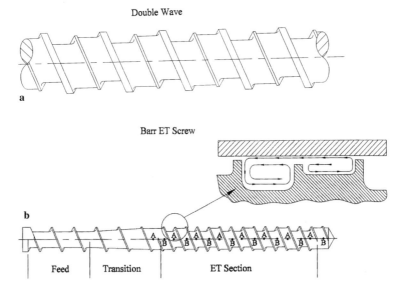

Figure 26 End of screw barrier mixers.

important implications. However, if the mixer jams—say from insufficient preheating and the screw breaks—this is not a very user friendly mixer.

Upstream Barrier Mixers. Given sufficiently large channels and barrier clearances, it is possible to move the barrier mixers further up the screw without jamming. The patent-pending Barr VBET® (Robert Barr, Inc) (Figure 27) is a long mixer extending to over one-half the screw length. Similar to the Barr ET® screw, this mixer has more mixing elements where the channel depth and radial clearance decrease as the material moves down

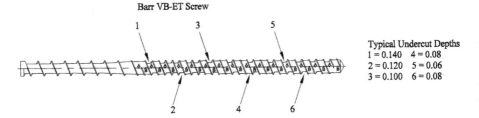

Figure 27 Mid-screw barrier mixer.

the screw. Because some material is driven forward and some pushed over the barrier, at least a small portion of the total flow will experience elongation.

The BT and Recirculator mixers can both be placed near the solids conveying zone (about the 10th L/D) without excessive pressure or jamming because of the open ended inlet channel. Up to four mixers are sequentially designed with increasing length and decreasing radial pump clearances (in response to the decreasing viscosity) in L/Ds of up to 50:1.

IV. SUMMARY

Single-screw extrusion for pharmaceutical is an orchestration of unit operations including solids transportation, melting, mixing, and pumping. The unit operations are considered separately, but it must be remembered that these do not exist in isolation. For example, the solids conveying rate must be matched to the pumping rate even though these occur at opposite ends of the screw.

While some parts of the extrusion process are quantified, much of the process is not. Texts do not, for example, give specific recommendations for a screw for any particular polymer—much less for a mixture of active drug and carrier. Because of the inherent secrecy involved in pharmaceuticals, it seems unlikely that a coherent text is viable.

Nevertheless, once the extrusion process is defined by the end user, the machinery is extremely reliable and the process readily controlled. With the increasing range available in biopolymers and tailored pharmaceuticals, growth in the use of single-screw extrusion will continue.

REFERENCES

1. Colby P. Plasticating Components Technology 2000. Youngstown, OH: Spirex.
2. Han KS, Spalding MA, Powers JR. Elimination of a restriction at the entrance of barrier flighted extruder screw sections. Plast Eng Mag.
3. Luker K. Multi-layer cast film and sheet systems for the lab. Film & Sheet Conference 97, Somerset, NJ, Dec 9, 1997:71–84.
4. Luker K. Single screw elongational mixing developments for continuous mixing applications. Medical/Pharmaceutical Extrusion Conference, Las Vegas, NV. Dec 5–7, 2001:111–136.
5. Luker K. Surge suppression—a new means to limit surging. SPE 54th Annual Technical Conference, 1996:445–451.
6. Luker K. U.S. Patent 5,569,429.

7. Rauwendaal C. Polymer Extrusion. NY: Hanser Publishers, 1986.
8. Rauwendaal C. Improve mixing by generating elongational flow. Continuous Compounding Conference, Beechwood, OH, Nov 14–15, 2000.
9. Song W. Extensional flow mixing technology for continuous compounding. Continuous Compounding Conference, Beechwood, OH, Nov 14–15, 2000: 323–351.
10. Tadmor Z, Imrich K. Engineering Principles of Plasticating Extrusion. Huntington, NY: Kreiger Publishing, 1978:221.
11. Thiele B. Twin-screw extruder theory for compounding, devolatilizing and direct product extrusion. Leistritz Twin Screw Extrusion Workshop, Somerset, NJ. June 11–12, 2002:6.
12. White J. Twin Screw Extrusion. NY: Hanser Publishers, 1990:42.

4

Twin-Screw Extrusion and Screw Design

William Thiele
American Leistritz Extruder Corporation, Somerville, New Jersey, USA

I. INTRODUCTION

For those unacquainted with twin-screw extruders (TSEs) there is cause for cautious optimism in knowing that the use of this well proven and versatile tool is within grasp to make a better, more repeatable, "good-manufacturing-practice" (GMP) product. However, given the peculiar nature of pharmaceutical-grade materials and processes, potential users should also exercise caution about importing solutions that worked for plastics and rubber, which served as the preamble for most of this work.

II. MOST BASIC PROPERTIES

This discussion will begin with the most basic properties of the key process elements for a twin-screw extruder, namely, the barrels, the screw elements, and the shafts on which they are staged and driven (1). Whether corotating or counterrotating, intermeshing or nonintermeshing, some basic statements may be made about these continuous, longitudinal, small-mass mixing devices (see Figure 1). Characteristics inherent with a twin-screw extruder include:

> TSEs are "sanitary." Designs can ensure that material is continuously exchanged on metal surfaces of the extruder to avoid stagnation.

69

Main Feed

Unit operations along the length of the extruder.

5 Mass Transfer Regions consisting of:

Pressure Output
- **Channels**
- **Lobal Capture**
- **Tip Acceleration**
- **Intermesh/Proximity**
- **Barrels' Apexes**

Figure 1 Continuous nature of twin-screw extruders.

TSEs are "continuous." A production run may continue uninterrupted until completion. This facilitates process stability and reproducibility.

TSEs are "small mass." The localized bounded domain of a mass of material is contained by the screw shapes and the barrels walls. Short, local, mass-transfer distances promote accurate distribution of small formulation constituents. In contrast, high-shear batch-mixer devices are often large mass because the processed material is bounded by the whole volume of the device. The transport distances to the moving elements and to the walls are comparatively large, causing accurate distribution of components to take more time. The short mass-transfer distances within screw sections promote both mixing accuracy and speed. Also, the heat transfer surface area in extruders tends to be about six times greater than that of batch mixers. This, plus the ability to engineer favorable heat transfer coefficients, helps to maintain critical temperature control. In general, "small mass" is a favorable property to maintain process control.

TSEs are "longitudinal." Over the length of the process section, consisting of barrels and screws, subprocesses (unit operations) may be sequenced as necessary to perform many of the manufacturing steps for the product. In fact, unit operations may be added and/or existing operations may be expanded until some boundary condition finally limits the system. Boundaries may include shaft torque, heat transfer, vent velocity, minimum dwell time, or several other limitations.

TSEs have "screw interaction." Both distributive and dispersive mixing may be enhanced by the flow patterns in the "apex region" where the two screws come together and in the "intermesh" or "proximity" region where the screws most intimately mesh and approach each other. Screw interaction is also used to enhance pumping and to separate subprocesses (zoning), as well as making twin screws more sanitary, or self-cleaning.

TSE processes receive "power through two screws." As a result of modern technology, shafts transmit high torque and can power a substantial length of screws, long enough for significant numbers of unit operations.

III. COMMERCIAL TWIN-SCREW EXTRUDERS

Medical/pharmaceutical extruders have been adapted from the food and plastic industries. The strongest, most widely disseminated, and versatile come from the latter. A brief summary of the available types of twin-screw extruders are described in Table 1.

Nonintermeshing twin-screw extruders are less common for mixing applications because of weaker screws interaction and being less self-cleaning. Low-speed, late-fusion, counterrotating twin-screw extruders often use on-piece barrels and screws. Lower screw speeds and shorter process lengths support the execution of fewer unit operations. The most likely candidates for pharmaceutical reacting, compounding, devolatilizing, granulating, and

Table 1 Commercial Twin-Screw Extruders

Generic type and origin		Sample builders
Counterrotating, intermeshing		
Slow speed	Profile heritage	Cincinnati-Milacron, Krauss-Maffei, etc.
High speed	Compounding heritage	Leistritz
Counterrotating, nonintermeshing (sometimes called "tangential")		
High speed	Compounding heritage	Welding Engineers, JSW
Corotating, intermeshing		
Low speed	Profile heritage	L.P. (Colombo), Windsor
High speed	Compounding heritage	Werner & Pfleiderer, Leistritz, etc.

purifying therefore tend to be adaptations of the higher-speed intermeshing machines in both corotation and counterrotation.

Regardless of extruder type, the machine must be validation-friendly, with certifiable GMP construction and documentation that will support the validation process. Also, beneath and within all of that stainless steel and special construction, the machine needs to be a strong, specially conceived twin-screw extruder that utilizes technologies that have been refined over the past century in a variety of demanding processes.

IV. TWIN-SCREW ELEMENTS—CLASSICAL AND NOVEL TYPES

The most common twin-screw element shapes that are described in literature are generally old, with the last core patents filed circa 1950 (2) (see Figure 2). For these original patents, the mathematical cross-sectional profiles were designed for close, constant, intermesh tracking of the screws into each other during rotation (3). In the case of the most widely used mode, intermeshing corotation, that intermesh is called "self-wiping." This tracking property entirely determined the mathematics that shaped the classical cross-sectional

CLASSICAL COUNTER-ROTATION

Rotational Intermesh Tracking + Power Through Two Shafts

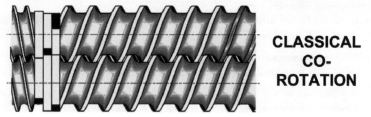

CLASSICAL CO-ROTATION

Figure 2   Classical intermeshing corotating and counterrotating screws.

profiles used for corotating extruders. Therefore, the cross-sectional profile shape had nothing to do with mixing, but rather with self-wiping.

The property of self-wiping was supposed to produce "self-cleaning." It usually does, but the primary jobs of extruders involve unit operations such as feeding, melting, mixing, draining, venting, and pumping, all of which were undertaken in classical twin-screw extruders under the constraints that the screw elements had to obey this intermesh tracking formula discipline. This is one of several reasons why technology that has proven acceptable for plastics may not be acceptable for executing unit operations in pharmaceutical processes, which may require different heat and mass transfer, and for which strong intermesh tracking may be undesirable.

Classical, tightly meshing screws, corotating or counterrotating, are preferred for many pharmaceutical processes, even though this use was never envisioned. These devices can enhance pumping, subprocess separation, as well as maximizing process sanitation by being self-wiping. However, it is important to note that even open meshing screws, which do not self-wipe, have higher mixing rates on their metal surfaces to maintain sanitation than mixing bowls, for example.

Sanitation, the continuous exchange of material present on a metal surface, is primarily a matter of achieving threshold levels of mass transfer for each material. At or above the critical threshold level the extruder (or even the batch-mixing bowl) will operate in a sanitary manner. Therefore, one need not be constrained to classical screw types whose self-wiping may be excessive

Co-rotating Screws for Carriers with Liquid Lubricating Component
Double Injection

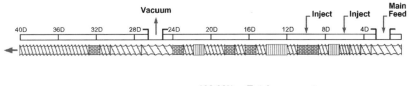

		100.00%	Total screw set
Classical mixing elements	140	14.14%	Kneaders
New distributive elements	180	18.18%	Combing (Forward & Reverse R&L)
New lobal dispersive elements	0	0.00%	n/a
Flighted elements	670	67.88%	Erdmenger std. bi-lobal

Figure 3 Mixed classical and newer screws.

Figure 4   Typical, nontraditional, corotating screws.

and may demand unnecessarily high energy from the screw shafts, at the expense of producing excessive mass or localized viscous heating.

Usually the optimum screw design will consist of a mixture of classical self-wiping screw types and new designs, which might not meet the threshold of being termed self-wiping (see Figure 3). Some of the "new" types are merely modifications of classical elements. Others are conceived solely to perform specific heat and mass-transfer operations (see Figure 4). In any case, screw types are chosen and sequenced along the screw shafts to perform the specific unit operations along the process length of the barrels/screws. Specific screw types will be discussed later.

V. DETERMINING SANITATION THRESHOLD LEVELS

In the plastics industry, the traditional market for twin-screw extruders, engineers tend to speak about "shear rate." This is unfortunate for two reasons. Strain rate usually consists of both shear and extensional components. Extensional mixing is often the most critical component, particularly for mixing low- and high-viscosity components together, which is frequently a requirement for pharmaceutical materials. Also, shear rate is not relevant to the resistance to straining. At the same shear rate, much more energy will be expended straining a higher-viscosity material, like tar, as compared to a lower-viscosity material, like water.

The sanitation threshold will, as implied above, be influenced by applied "stress rate." This stress rate is the product of the strain rate times the controlling modulus. The expression for this is as follows:

Basic Sources of Stress Rate

Formula:	$ds/dt = E_c \times de/dt$		
Nomenclature:	Stress rate	Controlling modulus	Strain rate
Relates to:	Work	Viscosities (resistance)	Movement
Units:	$(kg/m^2)/sec$	Pa sec/sec	(mm/sec)/mm
Applied:	$(kg/m^2)/sec$ = composite viscosity $\times$ strain rate $\times$ 0.102		
		(Pa sec)	("reciprocal seconds")
	(Divide by 1000 if your composite viscosity is in centipoise.)		

There are a couple of important issues to note. Strain rate in extruder screw channels may be about 100 sec^{-1} (+200/−90 sec^{-1}). Dispersive mixers can be thought of as about 1000 sec^{-1} (+2000 sec^{-1}, −700 sec^{-1}). Of course, the strain rate may be in shear or it may be in elongation.

The E_c ("controlling modulus") is the apparent viscosity of all the components together in their good or poor state of mix at that point in the extruder. For example, for compatible materials with viscosities that are different by perhaps a factor of three or less, the resulting E_c might be determined by their individual viscosities as proportioned in the formulation. If, however, a very low viscosity material is present at more than a few percent, it will disproportionately influence the overall E_c.

Taking these guidelines, water in screw channels would have a stress rate of only about 0.01 $(kg/m^2)/sec$. Polymeric composites might be about 10 to 1000 $(kg/m^2)/sec$, in the screw channels.

Interacting component viscosities that create the controlling modulus are not the sole contributors to sanitation threshold. Other factors include the surface chemistry of the extruder construction materials in combination with the resulting interactions with the processed materials. Materials that "stick" to the screw surfaces have elevated threshold levels. However, for any material there is a stress rate threshold for achieving process sanitation.

Generally, "sticky" materials have higher sanitation thresholds than watery ones. Again, for materials in any given morphological state it is

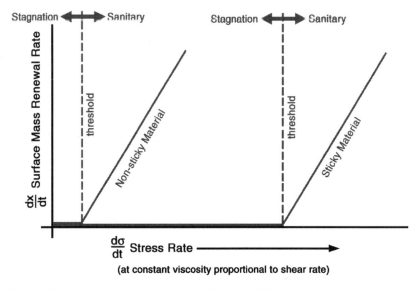

Figure 5   Sanitation stress-rate threshold for material types.

generally possible to determine a stress rate that makes the operation function in a sanitary manner. Should that stress rate be too high, solutions may include modifying the process sequencing, adjusting the formulation, changing the screws and barrels construction material, and/or a variety of other possible solutions (see Figure 5).

There are greater problems in maintaining sanitation in output adapters, and dies vs. the screws, because of the screws' higher stress rates. In adapters and connections, the self-cleaning screw properties do not apply. The design in this aspect of the installation is critical to maintain self-cleaning and sanitation. Cleaning is also an issue with wiped mixing bowls, but the relatively poor sanitation of bowls and other batch mixers makes the contrast with delivery and exit hardware seem less pronounced than with extruders.

VI.   GETTING STARTED

What is important in screws and barrels discipline for pharmaceutical applications is to:

Define the unit operations needed to be performed in the extruder to identify the candidate barrels configurations.

Take into account the required mass-transfer regions inside the extruder barrels profile for each unit operation, which relates to how screws behave inside the barrels.

Define, at least conceptually, the basic screw types.

Identify heat and mass-transfer requirements and boundaries for the unit operations.

Expect that a few iterations might be necessary, and that solutions and modeling for plastics applications may not completely apply. Most pharmaceutical applications involve customization of the extruder (4).

VII. BASIC SCREW TYPES

Rather than describing screw element types as either classical, by their common names or by special case, it seems useful to reduce the selection available to three basic functional types. Screw elements may be broadly classified as forwarding, mixing, and zoning (see Figure 6). It is not uncommon for a screw piece to have more than one of these attributes, and sometimes all three. Generally, however, a screw element will be mainly one of these basic types.

- **Forwarding**   **Almost Always Flighted**
  Feeding, Pumping, Driving Mixers

- **Mixing**   **Great Variety of Geometries**
  Dispersive and Distributive Mixing
  (Kneader shown)

- **Zoning**   **Restrictive Mixers and Flighted Elements**
  *Separates unit operations, assists
  mixers to function.*
  (Reverse mixer shown)

All screws in operation will perform one or more of these type functions.

Figure 6   Basic screw element types.

Forwarding elements perform the task that the classification implies and are used wherever there is an opening in the process, including at barrel holes to forward material away from feed openings, at vent openings to maintain a zero pressure, drain openings to positively convey materials, and at the discharge end of the extruder to pressurize the die. There can be exceptions, such as some injection holes to inject liquids or gases immediately prior to mixers.

Forwarding elements also serve as drivers to provide forwarding pressure to supply material into mixers. Some mixers are not self-centering in the extruder barrels and rely upon flighted elements and self-centering mixers and zoning elements to keep them properly centered. Forwarding elements usually perform this centering task (see Figure 7).

Forwarding elements are almost always flighted. The classical types are close meshing and self-wiping. The newer energy-focused elements are often open meshing, and not self-wiping in the traditional sense.

Mixing elements can be dispersive or distributive. Dispersive mixers are used to break down morphological units such as phase domains, droplets, and

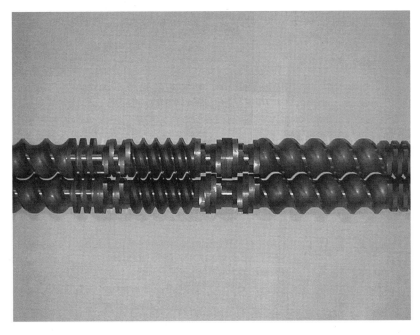

Figure 7   Driver-mixing sequence.

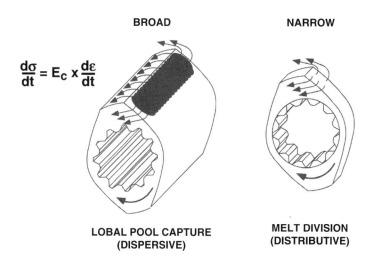

$$\frac{d\sigma}{dt} = E_c \times \frac{d\varepsilon}{dt}$$

BROAD

NARROW

LOBAL POOL CAPTURE
(DISPERSIVE)

MELT DIVISION
(DISTRIBUTIVE)

Wider disk = increased elongational acceleration/dispersive mixing
Narrower disk = melt divisions/distributive mixing

Figure 8   Basic kneading section for dispersive and distributive mixing.

agglomerates. Distributive mixers, on the other hand, space the morphological units without altering them. Similar to the case of sanitation stress-rate threshold, a stress-rate threshold exists for dispersive mixing (see Figure 8). Theoretically, no basic stress-rate threshold exists for distributive mixing (see Figure 9). Mixers may also be may be used to multiply surface area in devolatilizing and draining, and often serve to isolate unit operations from each as a "zoning" function (see Figure 10).

Besides being dispersive or distributive, a further subclassification may be added. Mixers may be forwarding, neutral, or reversing (see Figure 11). Forwarding mixers are capable of transporting material and may exhibit a small positive or negative pressure drop across the element. Neutral mixers have sections that push material neither forward nor backward, and may exhibit small to substantial pressure drops. Reversing mixers have elements that tend to push material backward and may exhibit moderate to large pressure drops. To most rules there are exceptions. For example, a forward-vaned gear or combing mixer may be productively paired with a reverse-vaned one to constitute a mixer pair that is both forwarding and reversing to perform more longitudinal homogenization (see Figure 12).

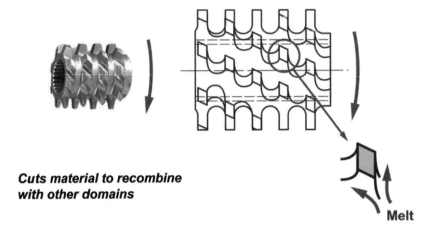

**Cuts material to recombine
with other domains**

Figure 9   Combing distributive mixer.

Dispersive mixers tend to capture material domains in pressure traps that cause the material to become squeezed, sheared, and elongated. The shapes of these mixers individually, and usually in interaction with the second screw, cause the processed material to become captured, deformed, and stressed when the screws turn (see Figure 13). Wide-section kneaders, lobal accelerators, and shearing discs are examples of dispersive mixing elements.

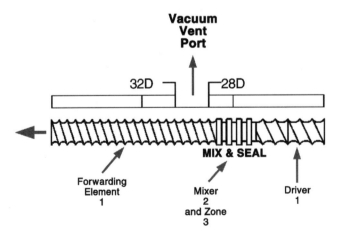

Figure 10   Mixer as a vent zoning element.

Kneading Blocks – co-rotation

30° Forward	90° Neutral	30° Reverse

Combing Mixers – co-rotation

Forward	Neutral	Reverse

Figure 11   Forward, neutral, and reverse combing elements and kneading blocks.

Combing Mixer	Combing Mixer	Combing Mixer
Vanes Opposed	**Vanes Aligned Left**	**Vanes Aligned Right**

Co-Rotation	**Co-Rotation**	**Co-Rotation**
Vanes push in in opposite directions *(circular)*	Both vanes push reverse	Both vanes push forward

Counterrotation	**Counterrotation**	**Counterrotation**
Both vanes push forward *(if left vanes, then in reverse)*	Vanes push in opposite directions *(circular)*	Vanes push in opposite directions *(circular)*

Figure 12   Opposing-flow distributive mixers.

1. Lobal capture element with acceleration shape
2. Lobal capture element with traditional shape
3. 0° Pitch lobal capture elements
4. 30° Forward close staged kneader
5. 30° Reverse close staged kneader
6. Shearing disc
7. 90° kneader with wide lobes

Figure 13 Lobal capture and shearing elements for dispersive mixing.

1 - 3. Combing mixer of increasing vane density
4. Lobal capture element with relief slots
5. Flighted element with distribution cuts
6. Disc with distribution channels

Figure 14 Sample elements for distributive mixing.

Distributive mixers tend to divide and recombine the material and ideally they do not capture locally pressurized domains like dispersive mixers do, but rather form easy paths for the dividing and recombining process to facilitate low energy and stress rate per division within the process section. Narrow section kneaders, gear elements, vane mixers, combing mixers, pin mixers, and interrupted screw flight elements are all examples of distributive mixing elements (see Figure 14).

Many common mixing elements have evolved to balance dispersive and distributive mixing (see Figure 15). Kneaders utilize the width of the sections and the angular advance or decline of the progressive stacked elements to set the strength of lobal capture, and dispersive mixing, by allowing controlled leakage paths for domain division, and distributive mixing.

The following table provides some basic generalizations for kneading sections for dispersive and distributive mixing.

Generalized Mixing Behavior of Kneading Blocks

Factor	Dispersive	Distributive
Section width	Wide	Narrow
Advance angle between sections	Small	Large
Direction of advance angle	Reverse	Forward

Zoning elements separate unit operations, e.g., to provide a pressure seal so that vacuum at a vent does not reach upstream or downstream to disturb other operations. Similarly, a zoning element might set the boundaries of an injected liquid or force a drain to release liquid (see Figure 16). Other zoning elements may block large particles from continuing down the screws following a melting or mixing zone. A restrictive zoning element might also be used to enhance the operation of a mixer, usually downstream of that mixer to ensure that it will be entirely full of material to best perform its mixing.

A pure zoning element provides no mixing itself, but rather serves a support function. A reverse-flighted screw element can be fairly pure by that definition as it is usually a poor mixer. Often mixers themselves serve "double duty" when is zoning need.

Examples of zoning elements include reverse-flighted elements, shearing discs, multisection neutral or reversing kneaders, and neutral/reversing distributive mixers. As a group, these screw elements serve some

	More Dispersive	More Distributive
Width:	Wider lobes	Narrower lobes
Staging:	30° angles between lobes	90° angles between lobes
Leakage:	Little bypass possibilities around lobes	Many bypass possibilities around lobes
	Much lobal capture	Reduced lobal capture
Type:	Dispersive with distributive component	Distributive with dispersive component

Figure 15   Dispersive elements with distributive leakage paths.

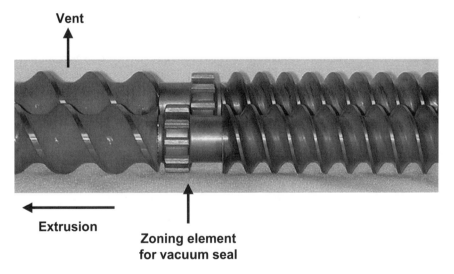

Vent

Extrusion

Zoning element
for vacuum seal

Figure 16   Zoning elements before vacuum vent.

kind of barrier function. Like with mixers, simple reasoning about the unit operations to be performed will provide a starting point for choosing zoning elements.

VIII.   MASS-TRANSFER REGIONS

One way to describe how a twin-screw extruder operates within barrels is to visualize the cross-sectional activity regions and the related affects inside the twin-screw extruder. Process designs may be created to promote or suppress as required activity in the five basic screw regions: channels, lobes, tips, apexes, and intermesh (see Figure 17).

Channels in a twin-screw extruder are comparatively gentle, mainly because the process melt is not captured and pressurized domains for heavy mixing do not occur. Most twin-screw extruders are starve-fed, where the feeders set the rate to the machine, and extruder rpm is independent. Therefore, the channels are not full, resulting in zero-pressure regions. While channel regions are generally associated with forwarding elements, they may also be designated in mixing and zoning elements. Processes dominated by channels tend to be less shear intensive relative to the other regions described below.

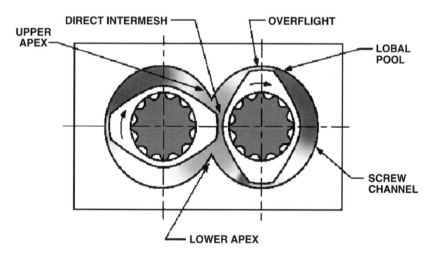

Figure 17   Five mass-transfer regions.

The lobal regions capture and pressurize material in "lobal pools," which are relieved up and over the screw tips to facilitate extensional and shear-strain rate components. Lobal capture powers most dispersive mixing processes. Typical kneading blocks are called "bilobal" because they capture two lobal pools. Deep flighted counterrotating machines may be up to hexalobal. It goes without saying that distributive processes avoid lobal events.

Screw tips usually provide shear and an exit for lobal pools with very high shear. For processes that cannot use shear stress for mixing, the shearing length and intensity should be minimized. Some lobal capture elements can be made to perform better as compared to the traditional kneaders. Extensional mixing is difficult estimate, but a reality.

Apexes are where the barrels join in lower and upper peninsulas of metal. These are screw regions that provide some moderate extensional and shear mixing. In corotation, the upper and lower apexes host a productive material direction change as the material transfers from screw to screw. Mild to moderate apex stress rates occurs to support distributive mixing and dispersive mixing of low stress-rate threshold. For the intermeshing, counterrotating twin-screw extruders, the lower apex provides a powerful lobal-like capture.

The intermesh is where the screws come together. While this region is often ignored in corotation, it hosts higher stress rates than the apexes, particularly with "sticky" materials. In counterrotation, it provides strong extensional and shear mixing components. For an intuitive understanding of the twin screw, the screw regions as described above are summarized below.

Generalized Processing Effects Within the Twin-Screw Five Mass-Transfer Regions

Region	Mode	Extensional mixing	Shear mixing
Channels	Co- and counterrotation	Weak	Weak
Lobes	Co- and counterrotation	Strong	Strong
Tips	Co- and counterrotation	Weak	Strong
Apexes	Corotation	Moderate	Moderate
	Classical counterrotation	Moderate++	Moderate
Intermesh	Corotation	Moderate to strong	Moderate to strong
	Classical counterrotation	Strong	Strong

Type of Processing:	Gentle	Strong
Screws' speed *Mixing rate in each of the five mass transfer regions at each point along the screws is proportional to screws' RPM*	Low	High
Screws' fill *Starving the feed increases remastication and time in mixers to increase process intensity.*	High	Low
Temperature *Lowering the temperature increases the controlling modulus in dispersive stress rate = E_c x strain rate*	High	Low
Screws' lobality *Increasing the number of lobal events and their intensity increases process severity. Within its boundary conditions the same is true of planar shear.*	Low	High
Sequential feed *Sequential feed may be employed to either capture or avoid high stress rate exposure.*	Depends	Depends

In twin screw extruder compounders, reactors and devolatilizers the balance between the rate determining feeders and the screws' speed is critical, not just screw speed, rate, temperature and screw design by themselves.

Figure 18   General adjustment for gentle and strong mixing intensity.

It is important to note that the feeding devices in twin-screw extruders set the rate, whereas the screw speed, screw design, and temperatures are used to control process severity in the unit. If the feeders are kept at constant rate, raising the screw speed will cause remastication in the high-stress regions, which will also occur at proportionately higher strain rates (see Figure 18).

IX. MIXING

In this chapter, so far, we have addressed dispersive and distributive mixing, as well as shear-stress rate and extensional-stress rate. This was in the context of screw element types, of mass-transfer regions formed by screws and barrels, and of achieving sanitation stress-rate level. However, the issues of pharmaceutical unit operations and mixing mechanisms to achieve target morphologies has not yet been discussed.

Pharmaceutical processes may also involve solids reduction, phase separations, precision distributions, reaction and purification operations, and validated reproducibility in a way more stringent than the world of plastics and food processing common to twin-screw extruders.

The basic model for distributive mixing is dividing and recombining material without disturbing the individual morphological components (see Figure 19). There is no stress-rate threshold, as no agglomerate, droplet, or the like will be reduced. The divisions, therefore, should be made as efficiently as possible. Using sharp leading edges to cut the processed material tends to be more efficient than dull cutting edges inherent with kneading elements, which require additional energy and cause viscous heating in the material and load on the screw shafts. Distributive divisions may be extended to cause ever more precise equal distribution of morphological components. In practice, twin-screw extruders arrive at functionally mature distributions rapidly. This is partly the product of being a small-mass device with short transport distances. Often distributive mixers alter morphological units and are effectively dispersive mixers, particularly when the material has a low threshold level for dispersive stress rate.

When the morphological components are to be reduced dispersive mixing is required. Extruders, mixing bowls, and most other mixers produce more shear strain, and therefore shear stress, than extensional strain, and extensional stress. Very much like the sanitation threshold, a certain critical stress rate must be achieved before the reduction process begins.

Most commonly, "shear" is thought of in connection to dispersive mixing (see Figure 20). In a shear field, the differential forces of the carrier acting upon a particle tend to cause a fissure in it.

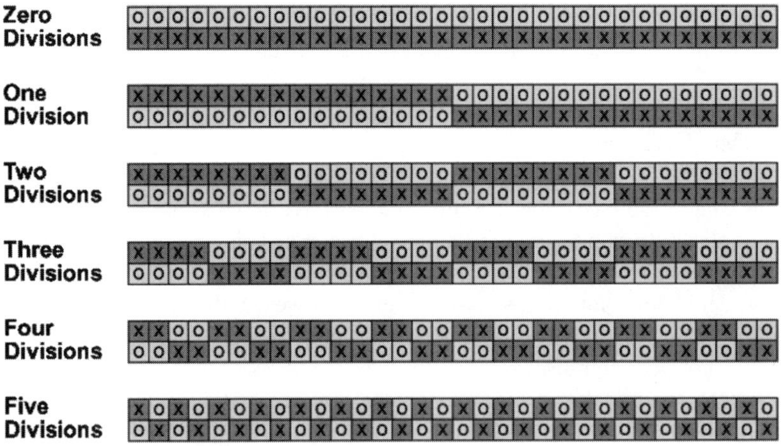

Figure 19   Distributive mixing model.

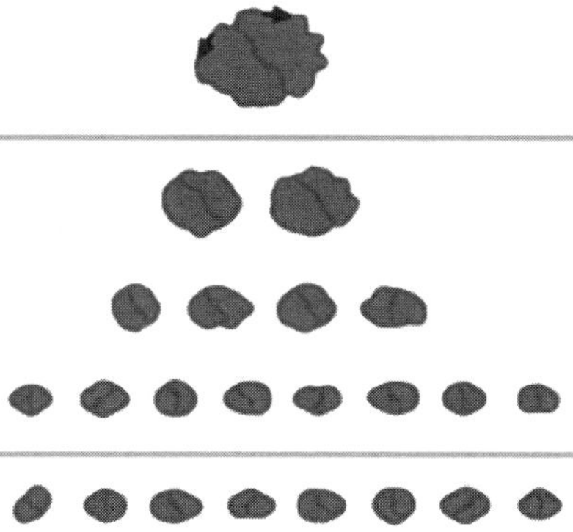

Division Number	Shear Rate Req'd	End Pieces	
1st	60	2	
2nd	240	4	
3rd	960	8	
4th	3840	8	
5th	15360	8	**Beyond Available Shear Rate**
6th	61440	8	

Figure 20   Shear mixing model.

Suppose a solids' agglomerate, in a mixing element, is in a shear-strain rate field of 1000 sec^{-1} and the agglomerate is low enough in concentration that the controlling modulus is governed by the viscosity of the carrier, which at that shear rate and temperature, from rheology data, is 100 Pa sec. The resultant shear-stress rate would be about 10,000 $(kg/m^2)/sec$. Actually that agglomerate only needed, say, 5000 $(kg/m^2)/sec$ to fracture into two or more pieces, so that division is easily accomplished. However, the half-sized particle remaining, which may still be too large, might require roughly four times the stress rate to break again because, in its reduced size, it is about four times as strong in a hydrodynamic shear field. Therefore, 20,000 $(kg/m^2)/sec$ shear-stress rate would be needed.

A few other things may also be happening. The carrier may be thinning from viscous heating. As higher screw speeds are attempted in hopes of higher stress rate, the result can be to reduce the viscosity because of shear thinning. The nature of the materials being processed must be considered relative to this point.

In practice, shear is used to successfully disperse agglomerates and droplets, often through many divisions. And for some morphologies, such as stacked platelets, it is the most productive mode, but with each size reduction, the smaller particles are significantly more difficult to break, and there are limitations.

Fortunately, extensional strain rates may also be generated in twin-screw extruders (5) (see Figure 21). As in shear-based strain rate, there is a threshold level for particle reduction. However, the reduced particles, and droplets in particular, do not seem to become as difficult to divide with decreasing mass, but respond to issues of surface chemistry, carrier viscosity and other values determining the ability of the particle to respond to the percent elongation of the field in which it is present during extensional mixing. What is known is that divisions of particles and droplets often may continue in extensional fields long after divisions would have stopped in shear fields.

When the stress rate, whether in shear or in elongation, falls below the necessary threshold level, dispersive mixing stops. Distributive mixing continues. Viscous heating continues. In order to recover dispersive activity the strain rate and/or the controlling modulus must be increased to raise the stress rate above the threshold for reducing the target morphological domain.

Screw elements with higher strain rates and running the screws at higher speeds might be beneficial. But these factors may also cause shear thinning and heating, which would reduce E_c, perhaps, more than offsetting the higher

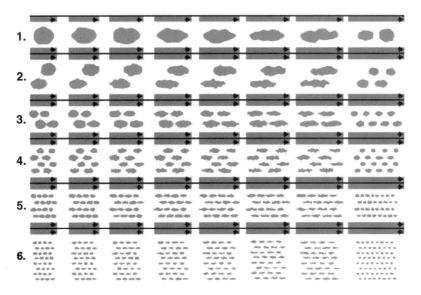

Figure 21 Extensional mixing model.

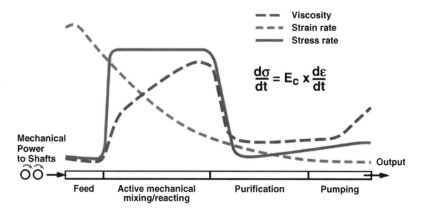

Figure 22   Stress-rate profiles down the length of the extruder.

strain rate (see Figure 22). Several possibilities exist to solve the problem. Using screws conducive to removing heat from the process via barrel cooling could decrease the temperature of the material, stiffening it, before mixing again. A formulation component with a lower, typically ambient temperature could be fed downstream into a melt to increase viscosity before mixing. Likewise, a liquid could be added and vented, carrying with it the heat of vaporization. Stronger mixers could be separated to mitigate viscous heating effects. The materials being processed determine the optimum approach.

There are cases where the threshold level is not a minimum, but rather a maximum, over which some unwanted action occurs. In such cases, some component might need protection from mechanical or molecular degradation caused by heat or stress. Gentle screws, downstream feeding where E_c has become very small, lubrication of the component, and special heat transfer barrels and screw elements are among the candidate solutions to these problems.

The differing demands of pharmaceutical processes preclude generalized solutions to screw and barrel design. Fortunately, twin-screw extruders can be modular so that customization of these devices is practical.

X.   SCALE-UP AND SCALE-DOWN

Many pharmaceutical products are initially developed via small batch processes, which might be impractical when scaled up as larger batch processes.

By plastics' industry standards pharmaceutical processes are small, making scale-up easier. Some boundary conditions and scale-up rules follow.

Volume is a boundary value. Each unit operation will use some of that volume. Extruders tend to scale volumetrically to about the cube of their screw diameters, provided their screw diameter-to-centerline spacing ratios are the same. Volume in each unit operation determines dwell time at a given rate. Many unit operations are greatly accelerated in the twin-screw vs. batch mixers.

Heat transfer is a boundary value. Each unit operation will need heat input or dissipation. Modern twin-screw extruders tend to scale thermally to about the square of the screw diameter. Twin-screw extruders may have about six times the heat transfer area to volume as compared to mixing bowls, and also have more favorable overall heat transfer coefficients per unit area of barrels. In many processes, most heat input originates from the main drive motor, through the gearbox, and from the screw shafts and screws, with rather nominal temperature-control requirements.

Mechanical energy is a boundary value. It originates at the motor, and is applied through the gears to the screw shafts. Both small and large state-of-the-art twin-screw extruders have similar power available to power each unit volume per rotation of their screws. It is this energy that primarily powers all the unit operations performed on the twin-screw extruder.

These and other boundary values generally determine the basic size of the extruder to produce a given output. Some expansion or contraction will be made according to specific scale-up situations. Most often, however, similar twin-screw extruders scale between the square and the cube of their screw sizes, depending upon whether the processes are more thermally or volumetrically dependent.

Distributive mixing is a direct scale-up value. The same divisions per kilogram in process A generally equals the same divisions per kilogram in process B. Scaling extruders can be entertaining. Even if the division rate of a mixer in a smaller machine is miscalculated, when miscalculated in the same way in the larger machine the result is usually valid.

Dispersive mixing is a direct scale-up value. The same product of time and stress rate at the same level over the threshold level in process A generally equals process B. Dispersive element groups in larger extruders can "look different" than in smaller machines. Scale-up values between sizes may require some empirical adjustments. Kneaders and other lobal capture elements dominate this group of mixers.

Coarse devolatilization is a direct scale-up value. Supply the same heat of vaporization per weight of volatile in process A as in B and the results will

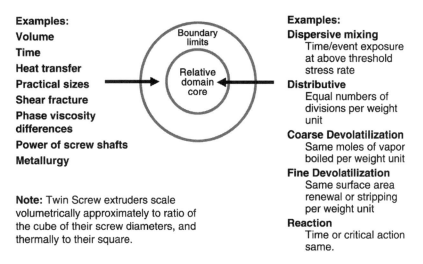

Examples:
Volume
Time
Heat transfer
Practical sizes
Shear fracture
Phase viscosity
differences
Power of screw shafts
Metallurgy

Note: Twin Screw extruders scale volumetrically approximately to ratio of the cube of their screw diameters, and thermally to their square.

Examples:
Dispersive mixing
Time/event exposure at above threshold stress rate
Distributive
Equal numbers of divisions per weight unit
Coarse Devolatilization
Same moles of vapor boiled per weight unit
Fine Devolatilization
Same surface area renewal or stripping per weight unit
Reaction
Time or critical action same.

Figure 23 Basic scale-up small vs. long machine.

generally be equivalent, having boiled off the same relative moles of volatile. Boundary conditions may include critical barrels' area per mole of volatile and vent velocity. Channels of simple flighted elements dominate this screw element type, which are sometimes supported by viscous-heating mixers.

Fine devolatilization is a direct scale-up value. Equivalent surface renewal with or without matched stripping yields scale-up. A variety of mixers and flighted screw elements are used.

Reaction dwell is a direct scale-up value. Given the same balance of activity in the mass-transfer regions and temperature, or trade-off balancing, scale-up is achieved. Almost all screw types and mass-transfer regions are used (see Figure 23).

Generalizations fail with nonstandard materials and processes. But if the process is divided into its unit operations within the extruder, it is generally possible to choose modular parts to service each subprocess, especially if that product is already running in a batch device or another continuous-mixing device.

XI. UNIT OPERATIONS

The unit operations in a typical plastics twin-screw extruder are feeding, melting, mixing, venting, and pumping (6) (see Figure 24). While some pharmaceutical processes are simple, some are not. Pharmaceutical twin-screw process designs tend to be considered proprietary and confidential. Therefore, the unit operations concept will be described with a merely hypothetical configuration.

The process could be simple mixing and pumping, or alternatively could be reaction steps, among other applications. This sample unit operation as described below involves preparing a carrier, to which active and inert components are added, a normal devolatilization step, a purification step, and sanitary discharge (see Figure 25).

Example of Possible Staging of Subprocesses in a Pharmaceutical Extruder

Unit operation	*Barrels/Screws*
1. Solids feeding of carrier components.	#1 Feeding, with nitrogen purge.
	Forwarding close meshing.
2. Semimelting and mixing of carrier components.	#2 Solid barrel section.
	Distributive/Dispersive mixers.
3. Near atmospheric venting of volatiles.	#3 Vent. Light draw-off and condenser.
	Forwarding close meshing.
4. Feeding active and inert solids.	#4 Side stuffed feed barrel.
	Zoning and forward close meshing.
5. Mixing solids into carrier. (Liquid could be injected here.)	#5, #6 Solid barrel sections.
	Distributive then dispersive.
6. Gas stripping for purification.	#7 Solid barrel with gas injector.
	Zoning, inject to distributive mixers.
	#8 Vent. Vacuum, Vent gas/impurities.
	Distributive mixers, forward close mesh.
7. High vacuum devolatilization	#9 Vent. High vacuum. Condenser.
	Zoning seal. Forward close meshing.
	#10 Solid.
	Forward open meshing to zoning seal.
8. Sanitary discharge and compactions.	#11 Solid.
	Forward open meshing to compaction.
	Material exits from divider/distributors.

Heat exchange separate to each operation

Figure 24   Unit operations of a simple plastic's compounding extruder.

By convention, the length of a typical twin-screw barrels section is four times the screw diameter. With 11 sections, this extruder would be 44 L/D long; 32 L/D to 44 L/D is the common range of lengths, 60 L/D being less typical. Each of the unit operations will be established according to its needs, such as divisions per kilogram, target dwell time, run length need from a feeding port, venting surface, product of time (lobal events) at or above threshold level in dispersive mixing, and so forth.

XII. SCREW AND BARREL DESIGNS FOR TOTAL SYSTEMS

Extruders may be surrounded by support devices such as feeders, prereactors, driers, pumps, condensers, bulk premixers, granulators, and many other pieces of equipment. Tandem extruders, where a primary extruder melt feeds a secondary extruder of a different design, are also possible. Multiple extruders can be linked in a system to make layered products by coextrusion. Other than the common stainless-steel-dominated GMP constructions, equipment configurations and applications differ greatly between pharmaceutical projects. Other chapters describe these variations.

Heat exchange separate to each operation

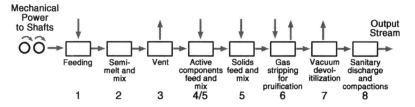

Figure 25   Unit operations of a hypothetical pharmaceutical process.

If the total system has the minimum number of necessary devices running well, process validation, operation, and clean-downs will be simplified. Part of this strategy may often include consolidating as many unit operations as possible on continuous, sanitary, twin-screw extruders. By analyzing the unit operations, process streamlining is often possible.

XIII. COROTATION OR COUNTERROTATION?

Corotating twin-screw extruders are generally the extruder of choice because they are available in the newest high-strength and heat transfer configurations. Additionally, corotators are excellent feeding devices for powders, pellets, and fibers. In this design, the intermesh wiping rate with close-meshing screws can be the highest of any extruder or batch device. This is important only when the sanitation stress-rate threshold is very high.

Corotators suffer in two areas: pumping is via drag flow, like single-screw extruders, and in "typical" flight depths, the lobe count of the screws is limited to two (see Figure 26).

Counterrotators solve these two problems. Only ram extruders and close meshing counterrotating extruders are non-drag-flow pumping. This is useful for managing high-percentage, low-viscosity phases. Because of a common inter-

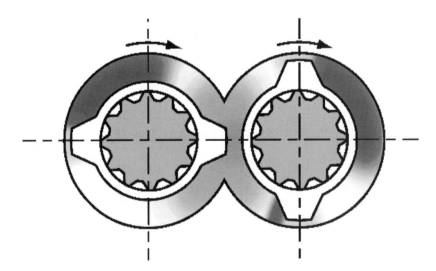

Figure 26 Cross section of bilobal mixer.

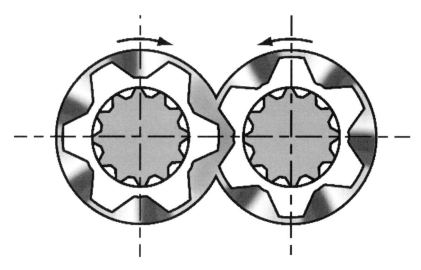

Figure 27 Hexalobal mixer cross section counterrotation only.

mesh movement direction in counterrotators, lobe counts of six and even eight may be incorporated to accelerate dispersive mixing (see Figure 27). State-of-the-art counterrotators need not rely upon intermesh mixing as their classical versions did, but rather upon lobal mixing to achieve mixing, just as in corotation (7).

Screw and barrel designs in corotation and counterrotation are basically the same. However, commercial realities heavily encourage opting for corotation for many applications.

XIV. SUMMARY

A generalized approach has been given to screw and barrel designs for pharmaceutical applications. Twin-screw extruders are sanitary, reproducible, continuous, small mass and longitudinal, and can utilize modular platforms upon which unit operations may be sequenced.

Pharmaceuticals are different than plastics. It is not surprising that different barrel configurations and screw types are used. Twin-screw components are chosen to perform unit operations of the total process. Efficiency is important to maximize productivity and inclusiveness of unit operations.

Twin screws are scale-up-friendly between each other and from most batch processes. As with any process, boundary conditions and scale-up values must be observed.

REFERENCES

1. Todd DB, ed. Plastics Compounding Equipment and Processing, Chapter 3. New York: Hanser, 1998.
2. Erdmenger R. German patents 815,641 and 813,154, filed Sept 1949.
3. Booy ML. Geometry of fully wiped twin-screw equipment. Polym Eng Sci 1978; 18:973.
4. Martin C. Continuous mixing/devolatilizing via twin screw extruders for drug delivery systems. Interphex Conference, Philadelphia, PA, March 20, 2001.
5. Utracki LA, Abdellah A. Compatibilization of Polymer Blends. Boucherville, PQ: Canadian National Research Council, May 1995.
6. Thiele WC. Trends and guidelines in devolatilization and reactive extrusion. National Plastics Exposition, Society of the Plastics Industry, Chicago, IL, June 18, 1997.
7. Todd DB, ed. Plastics Compounding Equipment and Processing, Chapter 3. New York: Hanser, 1998.

5
Die Design

John Perdikoulias
Compuplast International Inc., Ontario, Canada

Tom Dobbie
Porpoise Viscometers Ltd., Lancashire, England

Extrusion is a reasonably mature technology. The most demanding applications historically lie within the plastics industry from which great advancements have been made and from which pharmaceutical extrusion technology has evolved. The main job of the die is to give shape to the final product. Some extrusion processes are easy and the dies are simple, and some are not. Extrusion technology covers a variety of machine types, dies, and an even bigger multitude of materials. One of the most common problems is the mismatch between material, die, and machine. By reviewing some of the techniques in extrusion die design, and supplying a few insights, it is possible to design dies to produce quality products.

The die, the material, and the machine are all involved. At any stage (see Table 1) in the extrusion process, there is more than one flow type (e.g., at the entrance to the die system there is constrained flow, mixing flow, and bulk deformation. During a coextrusion there is also a free surface between the materials). Because of the historical difficulties with calculation, most engineers have focused on the quantitative analysis of pressures and flow rates based on simple formulae. With the evolution of computer-aided engineering, or CAE, software and computer models, detailed calculations can be made to see the fine details that lead to quality products.

The ram extruder is one of the simplest machines and provides good insights into what actually happens during the extrusion through a cylindrical die.

Table 1 The Four Distinct Stages and Five Flow Processes in an Extrusion Process

The four distinct stages in an extrusion process	
Melting and mixing	The material is made homogeneous in composition and temperature.
Transporting	The material is moved to the next process stage.
Shaping	Deformation of the melt into the final shape and orientation.
Finishing	Mainly the removal of heat but also shape control.
The five flow processes	
Low flow	Low stresses, e.g., surface tension, gravity, and stress relaxation.
Mixing flow	Dispersion, distribution, homogeneity, and work.
Constrained flow	Pressure driven and moving surface driven.
Free surface flow	Bulk deformation and surface deformation.
Bulk deformation	Variation with melting, solidifying, and pressure.

The material to be extruded is packed into a metal barrel with a cylindrical channel leading to a die. The metal barrel is heated. The barrel is set to some temperature. The material comes to thermal equilibrium with the barrel. (There is a thermal gradient across the barrel; typically less than 0.5°C.) The material is forced through a capillary die with a rate or pressure controlled plunger. The material exits the die and swells slightly (see Figure 1).

This technique is well suited to the precision extrusion of very high value materials. The material temperature can be controlled very precisely. The ram exerts modest and repeatable pressure on the material. This leads to minimal degradation of the extrudate as well as a very consistent extrudate diameter. Various improvements can be implemented to increase the efficiency of this simple batch process. Also, detailed understanding of material changes can be found through the use of simulation.

I. FLAT DIES

For the production of flat sheets or tape, as would typically be used for transdermal drug delivery systems, the material must be converted from the circular shape into a thin and wide sheet. If the sheet is not too wide (i.e., more like a tape) then the transition can be fairly simple as shown in Figure 2.

The balancing in this type of die needs to account for the shorter material path along the centerline versus the sides. If some correction is not

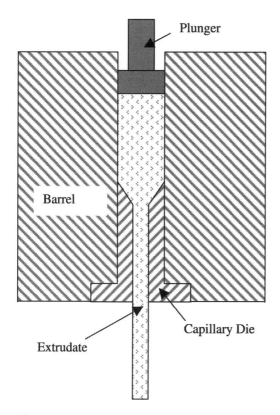

Figure 1 Simple ram extrusion using a capillary die.

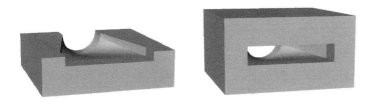

Figure 2 A simple circular to slit transition.

Figure 3   A slit die with a bow tie flow correction (left) and land correction (right).

applied, the sheet will end up being thicker in the middle than at the sides. One
method of correcting the shape is to enlarge the opening at the sides versus the
middle giving the cross section a "bow tie" appearance. Another method
would be to make the land section in the middle longer than at the edges.
These correction methods are demonstrated in Figure 3.

If much wider sheets are required, the die takes on a more traditional,
flat die shape shown in Figure 4. The various components for this type of
flat die system are comprised of a manifold, restrictor, relaxation chamber,
and die lips. These sections of the die are designed so as to achieve
a uniform distribution of material across the width with the shortest length
of die.

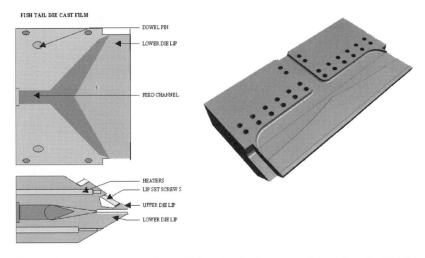

Figure 4   A typical "coat hanger" flat die (left) and an "Epoch" style (right).
(Courtesy of Cloeren Inc.)

In general, the design of the manifold and preland area is much more important than the lip gap adjustment method. The manifold and preland shape is what actually governs how well the die creates a uniform melt curtain. If this area is not properly sized and shaped for the material properties and throughput rate a nonuniform flow will result, and the operator will be unable to "fix" it using the lip adjustments.

The die lip adjustment capability should be thought of as being intended only for making very small changes to the material gauge thickness. Just as the take-off rolls are too late in the process to fix a nonuniform melt curtain, the die lips are too late in the die to fix a flow-distribution problem. This is a result of materials having a persistent "memory" effect.

II. ANNULAR DIES

The most common types of annular dies used in medical device applications for tubing are of a relatively simple "spider" type construction. Figure 5 shows a typical "spider" type annular die. The name comes from supports extending from the central mandrel to keep it centered to the body. The dies basically consist of a mandrel and body that form the main annular section and a set of die lips or tooling that form the final annular section near the exit of the die. These are also referred to as a central pin and outer bushing. The die lips are designed to form the material to the desired size prior to exiting the die.

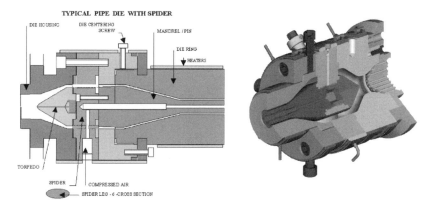

Figure 5   A typical spider die (left) and a 3-D view (right). (Courtesy of Genca.)

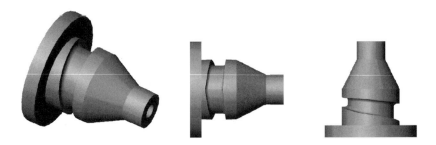

Figure 6 A typical side fed die.

Additional types of annular dies include "side fed" (or "cross-head") and "spiral mandrel." Examples of each of these are shown in Figures 6 and 7, respectively. Side fed dies are normally used in applications where a substrate needs to be coated because they allow a straight through path through the axis of the die. Spiral mandrel dies generally provide improved product uniformity because of the mixing effect of the spiral distribution

Figure 7 A spiral mandrel type die. (Courtesy of Genca.)

system and are commonly used in the production of polymer films. The majority of annular dies incorporate some sort of centering adjustments of the exit lips in order to adjust the thickness uniformity of the product. More detailed information on extrusion dies can be found in textbooks by Michaeli (1) and Rauwendaal (2).

III. DESIGN CRITERIA

In the description of the geometry of the various dies in the previous sections, the main design criterion that was described was the thickness uniformity. However, it must be emphasized that there are other equally important criteria that define an extrusion die. Probably the next most important criterion is residence time and residence time distribution. Simply put, this is the amount of time that the material spends within the extrusion system. This is especially important in pharmaceutical applications, as the materials are generally more prone to degradation when exposed to processing temperature for a prolonged time. The main reason that a spiral mandrel type of die may not be well suited for pharmaceutical material extrusion is that it is prone to imparting a large residence time distribution to the material.

There are basically two aspects to residence time with respect to die design: mechanical and rheological. From a mechanical aspect, the residence time is affected by how well the components that make up the die are machined and polished. Mating surfaces must match precisely and, where possible, should be machined and polished together. Any mismatch in mating surfaces can provide a "ledge" where material can stagnate and degrade. The mechanical aspect is relatively easy to observe and control, but the rheological aspect to residence time is a more difficult to control. This is because the residence time distribution depends on how the material flows within the channels and what shear stress is exerted on the flow channel walls. In order to emphasize this point, consider the flow of 1 cm^3/sec of material in tube with the 100-mm-diameter channel. This will result in an average velocity of about 0.127 mm/sec. This is of course an uncommon and extreme condition as the channel is much too large for this small flow rate, but it will help to demonstrate the point. It is relatively easy to imagine that no matter how smoothly polished the internal surface of the channel is, the average velocity of the material will be too slow, resulting in excessive residence time. Furthermore, the material will tend to define its' own smaller flow channel in the center of the large channel, leaving a large portion of essentially stagnant material near the walls. This condition is referred to as "channeling" and must be avoided.

Channeling occurs when the channel is essentially too large for the required flow rate. It is also affected by the rheological properties of the materials. What this means is that the flow channel needs to be sized so as to achieve sufficient shear stress on the wall to keep the material moving. The actual value of the shear stress depends on the material but can usually be easily determined from some simple experiments.

The next criterion that also needs to be considered is the overall resistance or pressure drop through the system. The pressure drop is related to the residence time in that a lower pressure drop or resistance generally implies a larger residence time (all other parameters being constant). In fact the pressure drop is the sum of the shear stress over the entire flow channel, which means that the higher the shear stress, the higher the pressure drop through the

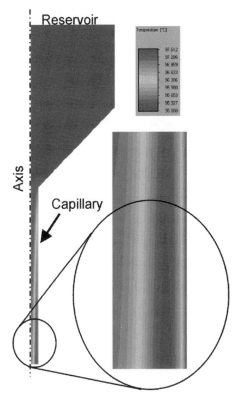

Figure 8   Temperature development in a capillary die (FLOW 2000™).

system. However, generally the designer's goal is to limit the maximum pressure drop through a die because the pressure drop energy is converted to heat within the die. This is referred to as viscous dissipation or shear heating. An estimate of the bulk temperature rise that results from the conversion of pressure drop to heat can be obtained from the following formula.

$$\Delta T = \frac{\Delta P}{\rho C_p}$$

where ΔT is the temperature rise, ΔP is the pressure drop, ρ is the melt density, and C_p is the heat capacity. However, the problem with the above formula is that it gives the theoretical, steady state, bulk temperature rise. Because of the low thermal conductivity of polymer melts, viscous heating is generally confined to the areas of high, local shear rate (near the wall). Since it is difficult to measure this effect precisely, engineers rely on CAE analysis tools to avoid these problems. This is demonstrated with the following example using the commercially available FLOW 2000™ CAE software (3).

Consider the flow in a simple capillary die shown in Figure 1 with a reservoir diameter of 10 mm and a capillary of 1 mm. The material temperature in the barrel is usually precisely controlled and in this example it will be assumed that the required temperature is 95 °C. Some typical material properties are selected (Cp=2000, k=0.15) and the flow conditions are simulated at piston speed of 0.5 mm/s. Figure 8 shows the development of the temperature

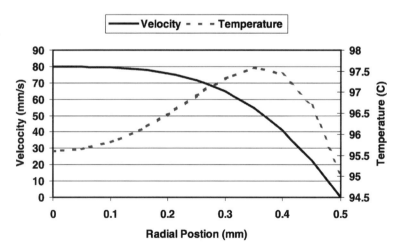

Figure 9   Velocity and temperature profiles near the capillary exit.

along the capillary. It can be seen that the temperature is not uniform and that the material is hotter near the wall. Figure 9 is a graph of the temperature vs. radial position in the capillary near the exit.

Figures 8 and 9 show that there is a maximum in the temperature a short distance from the wall, which is controlled at the desired temperature. The heat generation takes place at the high shear region near the wall and because of the low thermal conductivity of the material, the heat is not easily transferred away. This is an important issue in all extrusion processes and die design but it is especially important in medical device and pharmaceutical applications where the processing range (or window) of the materials is often much smaller.

IV. MATERIALS OF CONSTRUCTION

The materials for construction of a die for pharmaceutical applications need to be very stable and inert. For this reason, high-grade stainless steels that are resistant to corrosion are usually specified. The most resistant to corrosion are the 300 series, with 316 stainless being the most common. The drawback to this material is that it is relatively soft and can be damaged easily. Most die manufacturers prefer to use a 420 or a 17-4PH stainless steel, which have slightly less corrosion resistance but are harder and more durable materials. The higher temperature materials can also withstand oven cleaning; however, when doing so, care should be taken not to exceed the hardening temperature, which will result in annealing of the material. Capillary dies are often made out of tungsten carbide to ensure dimensional stability during cleaning in an oven.

Surface coatings like chrome or nickel plating are not generally used in pharmaceutical applications because the base steels are not generally corrosion resistant and there is a possibility that the coatings may flake off and contaminate the product. Records of material specification and quality assurance (QA) test should be maintained and are generally supplied with the dies.

V. MANUFACTURING

Dies for medical device and pharmaceutical applications tend to be very small and so the material costs are relatively low in comparison with the manufacturing cost. The materials may be stress relieved before and after machining and generally hardened after machining and polishing. Standard turning and milling equipment can be used for most dies, but some very small components and high precision may require more specialized techniques. One of these is the electrical discharge machining (EDM) method.

EDM uses a high voltage discharge to essentially vaporize the metal at the surface of the electrodes and thus form the desired shape by controlled erosion of the metal. The electrode can take the male form of the cavity that is desired and then plunged into the steel or the electrode can take the form of a wire that cuts through a plate in a similar fashion to a typical band saw, but much more precisely.

VI. SUMMARY

Almost every pharmaceutical extrusion system uses some type of forming die that results in the desired end product shape prior to sizing and cooling of the extrudate. This aspect of the extrusion system is no less critical than the material formulation, extruder type, or downstream system. A working knowledge of the material being processed, in combination with computer-aided design and working experience in "fine tuning" the internal flow geometry, are all important factors for producing a successful pharmaceutical extrusion die.

REFERENCES

1. Michaeli W. Extrusion Dies for Plastics and Rubber. Hanser Gardiner, 1992.
2. Rauwendaal C. Polymer Extrusion. Hanser Gardiner, 1994.
3. FLOW 2000 Suite of CAE Tools for Extrusion, Ver. 5.1, Compuplast International, 2001.

6

Material Handling and Feeder Technology

Winfried Doetsch
K-Tron Switzerland Ltd., Niederlenz, Switzerland

I. INTRODUCTION

Feeders are used in conjunction with extruders in the pharmaceutical industry for both melt extrusion and wet granulation processes to accurately meter solids and liquids. Feedstocks include pellets, powders, granules, gases, and liquids. Materials are either continuously metered directly into the extruder, or alternatively into a vessel to facilitate a premix that will then be fed into the extruder. Feed streams are introduced into the extruder either in a "starve-fed" manner, where the rate is set by the feeders and is independent of the extruder screw rpm; or alternatively via gravity, referred to as "flood feeding," where the hopper sits over the feed throat opening and the extruder screw rpm determines the throughput. Twin-screw extruders are typically starve fed, and single-screw extruders are normally flood fed. Figure 1 shows a typical, continuous, starve-fed melt extrusion installation with multiple feeders.

II. PREPARATION OF DRY AND WET INGREDIENTS

Ingredients are fed into the extruder in the correct ratio as defined by the recipe. The excipients and the active ingredients can be fed into a container in batch quantities. After charging, the container contents are mixed. This

Figure 1   Starve-fed continuous melt extrusion installation with multiple feeders.

procedure is called premix preparation. Afterward the premix is fed into the extruder via starve feeding or gravity, depending upon the extruder type. The alternative is to prepare all feedstocks individually and to convey each ingredient separately to designated feeders, which are mounted above the extruder.

A.   Premix Preparation

Process Description

Dry ingredients are supplied in drums, barrels, sacks, bins, bags, or containers and are usually discharged manually or automatically into containers. Figure 2 shows a typical automatic dispensing system.

After dispensing, the contents are often actively mixed (see Figure 3). The required mixing time is product and process dependent. After the mixing, the container is transferred to the extruder. The container is often used as a refill device for the feeder, which then meters the premix into the extruder.

Dispensing systems can weigh the dry ingredients in gain-in-weight batching (GWB) mode, in loss-in-weight batching (LWB) mode, or in loss-in-weight feeding (LWF) mode. The principles of each operation are explained below.

Figure 2 Automatic dispensing system.

Automatic Premix Production in Gain-in-Weight Batching Mode

Volumetric feeders are used in GWB systems. All dry ingredients are fed in sequence until the correct amount is reached. The container is positioned on a scale or on load cells (see Figure 4).

During the GWB cycle the weight of the vessel is continuously measured and is transmitted to a controller. Figure 5 shows the dependency between the delivered batch weight and the batch time. The weight before beginning the batch must be stable so that an accurate starting weight can be determined. When filling, the feeder starts to feed at a high speed, referred to as the "full-flow" mode. The weight increase during the batching is constantly compared to the setpoint. As the batch weight approaches the setpoint, the feeder reduces the filling rate to optimize batch accuracy, referred to as the "dribble-flow" mode. The feeder is stopped when the setpoint is reached. Upon completion of the batch, the scale is allowed to stabilize and the actual

Figure 3 Container mixer.

batch weight is recorded. A batch complete signal allows for the next ingredient to begin feeding. Most of the bulk solid is fed during the full-flow mode. The dribble cycle must be long enough and the feeder's speed must be low enough to minimize variations of the batch weight. Liquids may be added in a similar manner. The benefits of the GWB mode are simplicity—a single-scale system allows for simple wiring and control, and economy—lower investment costs because only one scale is used.

Automatic Premix Production in Loss-in-Weight Batching Mode

In LWB systems the feeders for the dry ingredients are mounted separately on load cells. In the case of simultaneous feeding, a certain degree of mixing occurs during the batching process and, therefore, the residence time in the

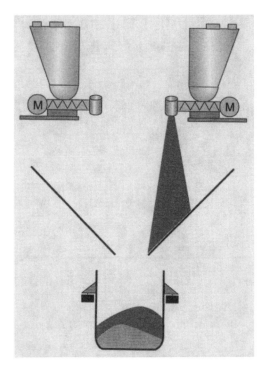

Figure 4 Gain-in-weight batching (GWB) system.

following mixing process can be reduced. The container into which the dry ingredients are fed is typically not positioned on a scale (see Figure 6).

During the LWB cycle the weight is measured and transmitted to the controller. The weight before starting the batch must be stable so that an accurate starting weight can be determined. The feeder then starts to feed in a full-flow mode. The weight decrease during the batching is constantly compared to the setpoint. As the batch weight comes close to the setpoint, the dribble-flow mode occurs and the feeder stops when the setpoint is reached. In LWB systems the ingredients are usually fed simultaneously (simultaneous LWB), but could also be operated sequentially (nonsimultaneous LWB).

The benefits of the simultaneous feeding of ingredients in the LWB mode versus the GWB mode are:

Economy—smaller scales can be used for minor dry ingredients.

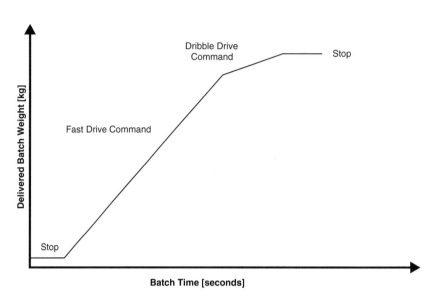

Figure 5 Steps of a GWB cycle.

Accuracy—higher feeding accuracy for minor dry ingredients because of increased scale resolution.

Rate—shorter batch time or smaller feeder sizes required.

Efficiency—premixing of ingredients occurs during the filling process.

Automatic Premix Production in Loss-in-Weight Feeding Mode

In LWF systems the feeders for the ingredients are mounted separately on load cells. All ingredients are simultaneously metered into the vessel with constant mass-flow rates according to the recipe. This procedure reduces the mixing time after dispensing to a minimum. The container into which the dry ingredients are fed is not necessarily positioned on a scale.

At short time intervals, the weight in the hopper is measured and transmitted to the controller. The current mass flow is calculated from the weight reduction per time. The screw rotation, if this is the feed mechanism, is continuously modified to compensate the difference between the setpoint and the current measured value of the mass flow.

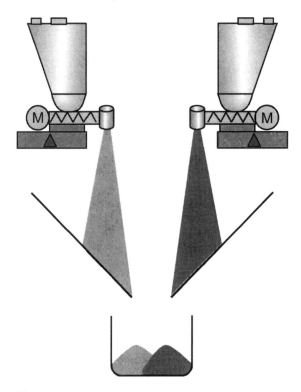

Figure 6 Simultaneous feeding in a loss-in-weight batching (LWB) system.

B. Dry-Ingredient Preparation

Process Description

For dry-ingredient preparation, designated discharge devices are used for each bulk pharmaceutical and excipient. The dry ingredients are discharged manually or automatically into containers or are transferred into designated pneumatic separators, which are used as refill devices for the feeders.

General Benefits of Dry-Ingredient Preparation

The benefits of dry-ingredient preparation versus the premix preparation are:

Smaller feeders can be used because of simultaneous operation.
Bulk batch blending and handling is avoided.
No intermediate storage of containers is necessary.

Segregation of different bulk solids in one container cannot occur.
Ease of production for drugs with different concentrations.

C. Liquid Feeding

Liquids are fed through a variety of pumps (e.g., gear, piston, or peristaltic pumps) with variable speed drive. The mass-flow rate can be measured and controlled by either a Coriolis mass flow meter with PID control or via liquid tanks situated on load cells with loss-in-weight (LWF) control. The benefits of a Coriolis mass flow meter versus a load cell arrangement are that no special refill devices are required, and the controls are relatively straightforward. The load cell arrangement as compared to a Coriolis mass flow meter is sometimes preferred because the system is easier to calibrate, there is no pressure drop caused by the measuring device, and the arrangement is suited for liquids with temperatures of more than 150°C.

III. FEEDER TYPES AND FEEDING MODE

A. Feeder Types

Screw Feeders

Screw feeders are presently the most widely used type of device for feeding dry ingredients into extruders. Either single-screw or corotating twin screws are specified. Other dry-ingredient feeder types include vibratory or belt feeders. The form of the material to be metered, in combination with the preferred mechanical design, determines the best choice.

For single-screw feeders, either auger or spiral screws can be used. Coarse or fine pitch screw profiles are specified depending on the desired mass-flow rate and the materials being fed. Twin-screw feeders can be used for feedstocks that exhibit poor flow properties, such as if the materials are floodable, compressible, adhesive, or cohesive. The "self-wiping" characteristics inherent with corotating twin screws assist in allowing for "first in, first out" mass flow in the hopper.

Screw feeders are manufactured to cGMP standards to meet pharmaceutical application requirements (see Figure 7). Pharmaceutical feeders are often characterized by the following:

All welds are sanitary and ground smooth and polished to match the base metal.
Machined stainless steel components are used instead of stainless steel cast parts wherever possible.

Figure 7 Screw feeders in cGMP format.

Figure 8 Disassembled horizontal feeder discharge piece with mesh insert for "lump reduction."

The feeder is easy to clean and disassemble.
Stainless steel shrouding is provided for the motor and gearbox (non-stainless-steel parts).
Validation documentation is available.

Modifications for Special Requirements

When feeding easily compactable bulk solids at low screw-rotation speeds (less than 30 rpm), the discharge rate is often not very uniform because the solid is discharged in lumps. For a more uniform discharge, a mesh can be fixed at the discharge outlet (see Figure 8). If melting or thermal decomposition occurs during conveyance, the screw feeder must be modified to reduce the friction, typically by changing the screw and sleeve clearances.

Aerated solids result in a very low bulk density that can actually behave like a liquid in the solids feeder. This can lead to problems if an empty hopper is refilled because the fluidized solids flood through the screws. Twin concave screw sets can help avoid this condition. Alternatively, a certain screw length is needed. Sometimes an isolation valve is required during the hopper refill. After reaching a certain level the isolation valve is opened (see Figure 9).

Figure 9 Twin-screw feeder with butterfly isolation valve.

B. Feeding Modes

Volumetric and gravimetric feeding modes are used for feeders. For reasons that will become obvious, gravimetric feeding is typically specified for cGMP installations.

Volumetric Feeding

A volumetric feeding device consists of the feeding module, the feeder hopper, a refill device, and a control system. The controller of the volumetric feeder sends a rotation speed signal to the motor. The rotation speed is sensed to facilitate closed loop control. The controller compares the rotation speed feedback with the setpoint in order to keep the speed constant at the setpoint. Constant speeds can result in high fluctuations in the mass-flow rate because of variations in material bulk density because of handling or head loads and/or variations in screw flight loading.

Volumetric screw feeders should generally not be used for metering dry ingredients into extruders for pharmaceutical applications because of the high fluctuations in the mass-flow rate and the difficulty in measuring and recording the mass-flow rate. However, volumetric feeders are often used as refill devices for gravimetric feeders.

Gravimetric, or Loss-in-Weight Feeding

A gravimetric feeding device consists of the feeding module, the feeder hopper, a refill device, a weighing device, and a control system. A loss-in-weight feeder estimates the mass-flow rate (quantity per time) by dividing the weight reduction by the time interval (see Figure 10).

The feeder is mounted on load cells. At short time intervals, the weight is measured and transmitted to the controller. The current mass flow is calculated from the weight reduction per time. To compensate for the difference between the setpoint and the current measured value of the mass flow, the motor rotation speed is continuously modified (see Figure 11).

The LWF mode is characterized by the following:

Gravimetric control is used to keep the mass-flow rate constant.

A measuring device monitors and records the mass-flow rate.

The screw-rotation speed (or other delivery mechanism) is adjusted to keep the mass-flow rate constant.

Relatively low fluctuations in the mass-flow rate if the feeder is mechanically correct.

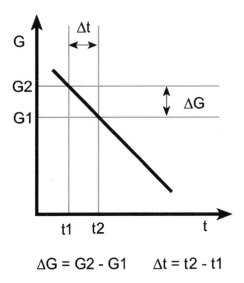

$$\Delta G = G2 - G1 \qquad \Delta t = t2 - t1$$

Figure 10 Weight reduction per time for gravimetric feeder.

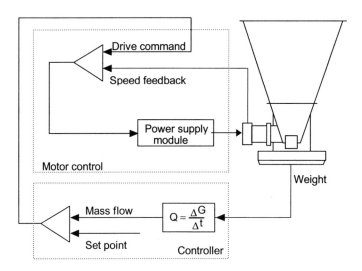

Figure 11 Control loop of a loss-in-weight screw feeder.

IV. FEEDER HOPPER

A. Design Problems

The design of the feeder hopper is strongly influenced by the flow properties of the bulk solids fed into the extruder. If the feeder hopper wall is too inclined versus the vertical axis or the surface finish of the feeder hopper wall is "rough," the velocity of the bulk solid is nearly zero close to the wall and higher farther away from the wall, which results in "funnel flow." This behavior is not acceptable for pharmaceutical extrusion technology, where mass flow is preferred. Mass flow is characterized by a movement of all particles, which are at the same height in a feeder hopper, with the same velocity downward for first in, first out mass flow. By decreasing the incline of the feeder hopper wall versus the vertical axis or by reducing the surface roughness of the hopper wall, mass flow can be realized. For bulk solids that tend to bridge, a driven agitator may be used within the hopper as an additional discharge aid.

B. Hopper Volume Considerations

The materials in a feeder hopper are strongly influenced by the volume flow rate. If a change in a refill device (e.g., container) takes a certain time, the volume, which is fed during this time, must be taken into consideration. A low level or "heel" must always be maintained within the hopper to avoid feeding inaccuracy.

Large feeder hoppers have the following benefits:

Lower number of refills.
More time to react and higher safety in case of refill problems.
Potential to deaerate bulk solids that fluidize.
Larger refill nozzles for poorly flowing bulk solids.
Higher feeding accuracy during the refill period is possible.

Small feeder hoppers have the following benefits:

Lower weight capacity of the load cells is required.
Higher weight resolution is possible.
Feeding accuracy is improved during the LWF period.
Less space and height is needed.
Possibility of feeding (with a cylindrical design) very poor flowing bulk
    solids without hopper agitation.

C. Mass Flow

Mass flow is important in feeder hoppers because "first in, first out" is realized and the residence time distribution of the bulk solid is very narrow. In rectangular feeder hoppers, mass flow is less apt to occur because of stagnation in the corners. Mass flow can also not be realized in asymmetrical conical feeder hoppers because the velocity of the particles at vertical and inclined walls will differ. Nevertheless, asymmetrical conical feeder hoppers are often used because a higher feeding accuracy is attainable if shorter screws are used, and a compact arrangement of several feeders in a circle is possible. Mass flow is most easily achieved in a cylindrical feeder hopper or in a symmetrical, conical feeder hopper. It is important to note that flow properties of bulk solids can be strongly influenced by environmental conditions, such as humidity, air temperature, bulk solid humidity, and bulk solid temperature.

D. Avoiding Bridging

Bulk solids can bridge in conical feeder hoppers in the zones where the cross section falls below the critical bridging diameter, particularly if no agitation

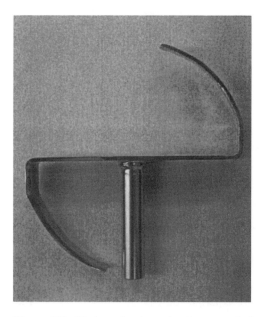

Figure 12 Horizontal agitator in pharmaceutical execution.

Figure 13 Twin-screw feeder with vertical agitator and enclosed vertical agitator gearbox.

device is used. Both horizontal and vertical agitators may be used (see Figures 12 and 13). Discharge aids help to guarantee a more uniform bulk-solid flow, but can reduce the feeding accuracy because of the resulting noise that can effect weight signals.

V. REFILL

A. General Statements

Because materials are continuously discharged from a feeder, a periodic refill is required. During the refill period, it is not possible to operate in the LWF mode because the weight increases. The time for refill should be short to reduce the refill time. During the refill, a defined volume of bulk solid is fed into the feeder hopper and the same amount of air is dislocated. The handling of the dislocated air and the usable refill devices are important factors to be considered to ensure metering accuracy.

B. Refill Levels

The loss-in-weight controller starts a refill when the minimum refill weight is reached (see Figure 14). The refill is stopped when the maximum refill weight is recorded. The refill weights are individually calculated for each bulk solid on the basis of bulk density. Changes in the bulk density influence the filling degree of the feeder hopper.

Loss-in-weight feeders typically discharge until the hopper is 60% filled and refill to 80% of the feeder hopper volume. The refill mass-flow rate is usually at least 10 times higher than the mass-flow rate of the feeder and about 10 sec in duration.

C. Refill Algorithm

The refill process starts when the minimum weight in the feed hopper is reached. The weight reduction cannot be measured during refill. Therefore, the

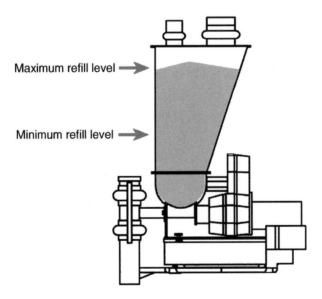

Figure 14   Minimum and maximum refill level.

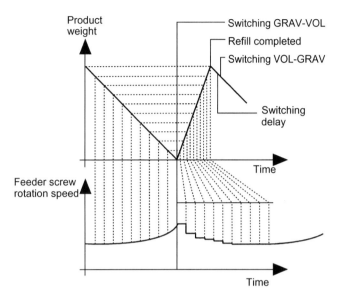

Figure 15 Refill algorithm for loss-in-weight feeder.

controller switches to volumetric control. Modern algorithms change the screw-rotation speed in the volumetric feeding period to increase the feeding accuracy during this period (see Figure 15).

The refill algorithm decreases the feeder speed in steps according to the information from the last gravimetric feeding cycle. The information in the registers defines the ratio between the feeder net weight (product weight) and the mean screw-rotation speed, which is necessary to achieve a defined mass-flow rate. When the refill maximum weight is attained, the feeder switches back to gravimetric mode after a short delay. The maximum and minimum refill weights, the refill mass-flow rate, the refill time, and the number of refills per hour all have to be adapted to the bulk solid properties and the operating environment.

D. Air Displacement

During refill the bulk solids displace air in the hopper. In the case of insufficient air venting, pressure is built up in the hopper, which can lead to an increased mass-flow rate during the refill time. Because the vented air contains dust particles, an exhaust system is required to reduce the dust emission.

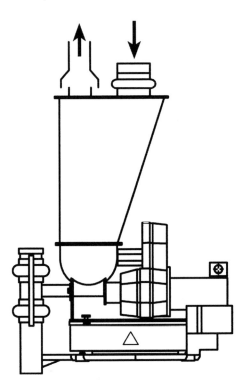

Figure 16 Suction of vented air in loss-in-weight feeder.

The simplest way of handling of displaced air is to weld a ventilation nozzle onto the lid of the feeder, or to drill a hole into the lid of the feeder hopper. At a relatively short distance from the hole, a suction tube is mounted (see Figure 16). The vented air is suctioned off by an exhaust air fan and is dedusted in an air-filtration system. The same arrangement may also be necessary between the feeder and the extruder to allow for back degassing by integrating a ventilation nozzle into the connection pipe between the feeder and the extruder.

E. Refill Devices

The design of the refill arrangement is important to obtain optimum feeding accuracy. Refill devices that are used include automatic refill devices, containers with tight isolation valves, or pneumatic refill devices.

If the refill volume is less than the volume of bulk solids in the container or in the pneumatic receiver, refill valves are required to avoid overfilling. The refill valve should be tight to avoid leakage during gravimetric feeding and placed close to the feeder for quick cut-off to avoid overfilling.

VI. WEIGHING DEVICE

A. Weight Balancing

It is important to postion the feeder in a way that no external forces or friction can influence the weighing. Flexible connections are generally required as connection elements to all other parts of the plant (see Figure 17). The following connections are used: conventional bellows fixed with band clamps or triclamp bellows fixed with clamps (see Figure 18).

Touching of the gravimetric feeding system by other plant parts or by persons has to be avoided. Cable or flexible tube connections for the supply of

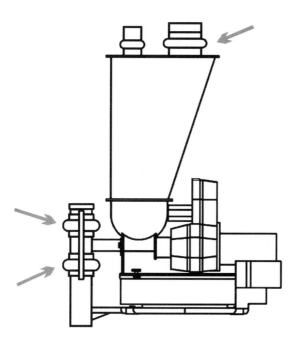

Figure 17   Flexible connection positions.

Figure 18 Triclamp bellows.

pneumatic drives or cleaning nozzles must be short and not stretched to avoid
external forces influencing the weigh scale.

B. Influences of Pressure Changes

Pressure changes in contained installations can lead to incorrect weight
readings and inaccurate feeding results. For example, a feeder with a closed
horizontal discharge and a refill nozzle with 100-mm diameter is mounted on a
scale and connected via one bellows with a lid. If the pressure in the feeder is
increased or decreased by only 1 mbar, the weight changes by 80 g.

This can be easily explained. A nozzle with 100-mm diameter has an
area of 0.00785 m^2. A pressure change of 1 mbar is equivalent to 100 N/m^2.
By multiplication of the pressure change with the area, a force of 0.785 N is
calculated, which is equivalent to 80 g.

If a pressure change of 1 mbar takes place within 10 sec in the described installation, a mass-flow rate of 8 g/sec (28.8 kg/hr) will be displayed without feeding any bulk solids. This shows how important pressure changes are, especially at low mass-flow rates. To eliminate the influence of pressure changes on the weighing results, outlet pressure compensation devices and inlet pressure compensation devices are used.

The outlet pressure compensation device is required if a vertical outlet is used and pressure changes in the downstream equipment can be detected. By fixing the parts above the upper bellow and the parts below the lower bellow with unweighed parts, identical forces work in opposite directions and independently of the existing pressure. These forces cancel out each other at every pressure, referred to as outlet pressure compensation, so that the weighing result is not influenced by pressure changes.

During refill, the bulk solid displaces air. If the pressure changes because of the refill, the pressure change on the weighing result has to be eliminated by an inlet pressure compensation device. A connection pipe (pressure-equalizing line) between the inlet and outlet is recommended to avoid a differential pressure between the inlet and outlet (see Figures 19 and 20).

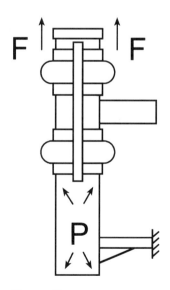

Figure 19 Outlet pressure compensation device.

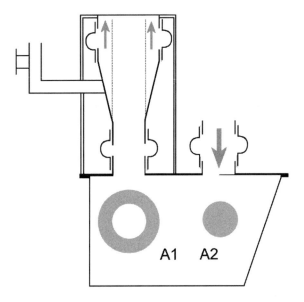

Figure 20   Inlet pressure compensation device.

C.   Influences of Vibrations

Modern load cells and control algorithms can eliminate the effect of plant vi-brations. Vibrations can be generated, for example, by motors or people moving on support structures. The elimination of the effect of vibrations is realized by the use of special, digital load cells. To check that these control algorithms do not reduce feeding accuracies, small weights identical to the minimum resolution should be laid on the load cell and the weight displayed denoted.

D.   Resolution of the Load Cells

Load cells must have a high resolution for low mass-flow rates. For example, a typical scale with an analog load cell for feeding bulk solids at low mass-flow rates has a weighing range of 32 kg and a resolution of 1:65,000. This scale can detect approximately 500-mg weight changes. From experience it is known that the weight change per second should be five times higher than the minimum resolution, which is equivalent to 2.5 g/sec (9 kg/hr). When feeding at this rate, it takes the controller 20 sec to detect whether the setpoint is reached with a deviation of +1%, or not. When the mass-flow rate is reduced below 9 kg/hr, it becomes more difficult to detect improvements of gravimetric control against volumetric control. That means that the mass-flow rate should be more than 9 kg/hr to see a significant improvement of the gravimetric control.

A typical platform scale with a digital load cell for feeding bulk solids at low mass-flow rates with a weighing range of 24 kg and a resolution of 1:1,000,000 can detect 24 mg. The limit value can thus be reduced from 9 kg/hr to approximately 0.4 kg/hr. Digital load cells are necessary for low mass-flow rates. Feeding accuracy is also improved at higher mass-flow rates because of the higher resolution.

VII. SUMMARY

Depending on the process formulation requirements, dry and/or liquid ingredients are metered into an extruder either individually or as a premix. The feeder hoppers have to be designed properly to have a sufficient capacity to realize mass flow and to avoid bridging. It is also important that the correct design of the refill system be used. The weighing device and controller are critical for high-accuracy feeding. To achieve optimum accuracy correct mounting, elimination of weight changes caused by pressure changes, vibrations, and other environmental influences, as well as the use of high-resolution load cells are necessary. For pharmaceutical applications, only feeders meeting cGMP guidelines should be used.

REFERENCES

1. United States Food and Drug Administration. Current Good Manufacturing Practices. 21CFR, Part 210 and Part 211.
2. Wilson DH. Top Ten Frequently Asked Questions on Feeder Accuracy. Training materials for K-Tron America, New Jersey, 2001.
3. Boilard TG. Feeder selection for pharmaceutical extruder applications. Medical/Pharmaceutical Extrusion Conference. Technical paper delivered Dec 2001.
4. Billings JP, Bradley RO, Damon RH Jr, Hejzlar S, Nielson HA, Paelian O, Platt FL, Quinn D, Shapiro BH. Terms and Definitions for the Weighing Industry. Washington, DC: Scale Manufacturer's Association, Inc., 1975.
5. Wilson DH. Feeding Technology for Plastics Processing. Munich: Carl Hanser Verlag, 1998.
6. Willis F, Keegan P. Sanitary feeders minimize product contamination. Chem Eng 1999.
7. Winski J. Addressing Feeding Performance. Chem Process Sept 2001.

7

Rheology and Torque Rheometers

Scott T. Martin
Thermo Electron Material Characterization, Madison, Wisconsin, USA

I. INTRODUCTION

The characterization of molten materials for pharmaceutical extrusion processes is an important facet in many stages of product development. This characterization can be in the form of specific (absolute) rheological determination of an expected phenomenon (flow, temperature dependence, etc.) or a general (relative) "rating" of the system in comparative terms (expected degradation, product inspection, resistance to process, etc.). An extremely effective tool that has found application in both of these genres of testing is the torque rheometer system. Although traditionally used for plastic and food processes, this device has been embraced by the pharmaceutical industry to develop and produce wide-ranging formulations for solid dosage forms.

A torque rheometer is an instrument that is widely used in studying formulations, developing pharmaceutical dosage forms, and characterizing polymer (for example) flow behavior by measuring viscosity-related torque caused by the resistance of the material to the shearing action of the melting process. Its fundamental usefulness relates to the versatility of the instrument to simulate development or manufacturing processes within a small-scale "laboratory environment."

Before describing the system itself, it is important to define the primary characterization parameter obtained through the use of the instrument—torque. Torque can be described as the effectiveness of a force to produce rotation, and its mathematical function would equal the product of the force applied and the perpendicular distance from the line of action to the instantaneous center of rotation. By definition, torque can be described as:

$$M = F \times r$$

where M is the torque, F is the force applied, and r is the radial arm (distance from center of rotation to force applied).

Units of torque are typically given in [m g], [N m], or [ft-lb] (see Figure 1).

The usefulness of this measurement in the characterization of a material and/or process lies in the comparative analysis of the torque registered by the system. This gives the user a "measure" of the resistance of the material to the process (and conditions) at hand. Through methodical variation of these parameters—materials and processes—a thorough comparison of materials and a potential for optimization of the process can be achieved. This can lead to useful functionality for the user in many areas of testing, ranging from standard quality control to pilot-scale investigations and production processing.

Torque rheometers have been in existence for nearly a century, originally finding application within the baking industry (flour dough quality control). Although the configurations have experienced several evolutions—purely mechanical, electronic, microprocessor controlled—the

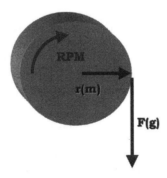

M = F x r, where...

M = torque

F = force applied

r = radial arm (distance from center of rotation to force applied)

Figure 1 Definition of torque.

basic goal of materials/process characterization has been well preserved (see Figure 2). What has been learned is currently being applied to develop GMP products.

The torque rheometer system has been developed in primarily a "building-block" fashion that allows the combination of different types of instruments to be mated to the torque rheometer, which is utilized as the instrument's main-drive motor. These instrument additions are in actuality "simulators" of the varying processes, and include miniaturized mixers (batch compounders), twin-screw extruders (continuous compounders), single-screw extruders (product processing instruments), and a host of auxiliary equipment to complete the "simulation" (see Figure 3).

It is important to note that although the torque rheometer is used as the main drive, this is not its primary function. An accurate and application-specific drive is important for the system; however, the torque rheometer's true utility is derived from its ability to process not only all of the signals from the sensors (temperature, pressure, feeding, ancillary conditions, etc.) but also primarily to maintain precise monitoring of the torque on the system. This is accomplished with a torque transducer (or load cell) positioned in line with the

Mechanical Torque Rheometer

The mechanical torque rheometer system was truly that – it relied upon a lever arm, weight, and dashpot configuration (dynamometer) to physically sense deflection due to motor resistance. The system proved quite useful for testing purposes, however due to its mechanical nature calibration was difficult and frequently required. In addition, numerous external 'modules' were necessary for the control/monitoring of parameters such as temperature and pressure.

Electronic Torque Rheometer

The next generation of torque rheometers were built primarily to address the inconveniences of the mechanical. The system was semi-automatic, relatively compact with most 'modules' and features (calibration, analysis) built in, and featured a unique measuring system that introduced a transducer to monitor axial movement of a lever arm/idler gear configuration.

Microprocessor-Controlled Torque Rheometer

This most recent evolution of the torque rheometer provides a complete incorporation of all facets of testing within one versatile, compact system. It allows for the user to not only control but also monitor, collect, and analyze information from a variety of signal sources. In addition, the system offers a variety of range-specific strain gauges, positioned directly in-line to the output shaft which thus produces a true torque signal with excellent resolution.

Figure 2 The evolution of the torque rheometer.

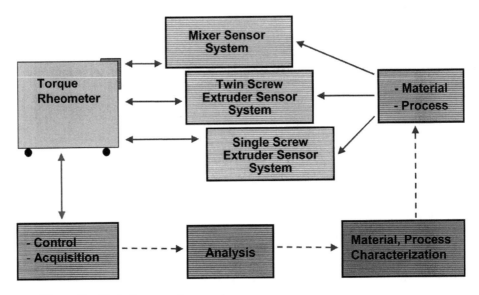

Figure 3 Block diagram of torque rheometer process.

attached sensor. An added feature of modern torque rheometer systems is an easily removable load cell (see Figure 4), which allows for application-specific matching of the load cell to the required testing range (to obtain better resolution in the torque signal).

A walk-through procedure applicable to generalized torque rheometry testing would be as follows:

1. A system (or sensor) is defined, based upon the objective of the process simulation.
2. Process conditions for initial testing are defined (materials, temperatures, instrument speed, etc.). It is noted "initial" because the modification of these to review the response/outcome is commonly the objective of the testing.
3. Testing is performed and optimized (if necessary).
4. Information through testing is collected. This information can include not only the testing parameters and variables (torque, set temperatures, resultant materials temperatures, resultant materials pressures, etc.) but also the material in its final form.
5. Analysis is performed based upon the results.

**Torque Rheometer
Base Drive Unit** *Motor*

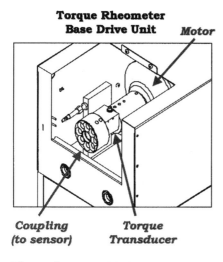

**Torque Transducer
Specification**

- **bonded strain gauge**
- **accuracy 0.1%**
- **resolution 16 bit**
- **reproducibility 0.02%**
- **interchangeable**
- **coded for auto detection**

*Coupling
(to sensor)*   *Torque
Transducer*

Figure 4   Removable load cell.

The typical output objective for testing is material and/or process characterization. This can be broken down into four areas of particular torque rheometry interest. These would be as follows:

Viscosity—the internal resistance of a fluid to flow under a defined shear rate and resultant shear stress.

Elasticity—a property of a material by virtue of which it tends to recover its original size and shape following deformation.

Shear sensitivity—the nature of a material to break down structurally when exposed to shear stresses (during processing).

Temperature sensitivity—the response of a material to its thermal history (during processing).

As an example of a common torque rheometer experiment and the results obtained, please refer to Figure 5. The graph, also known as a rheograph (torque vs. time curve), shows the response of a material to a process simulated by a torque rheometer/batch-mixer system. Torque and melt temperature are critical parameters to provide insight into the viability of the formulation being evaluated.

An evaluation technique quite common to rheograph data includes the observation of a torque signal at significant points (or times) during the

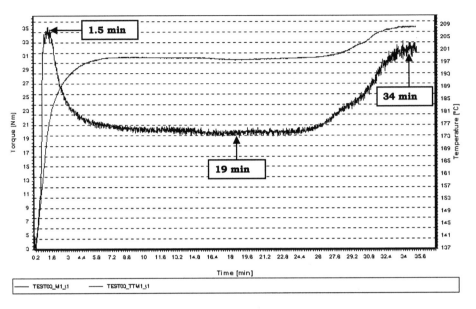

Figure 5   Batch mixer rheograph example.

testing. This data can be used for comparison purposes with a standard
material, to indicate physical or chemical state changes, or to determine the
work required in processing the material.

II.   TORQUE RHEOMETER WITH BATCH-MIXING SENSOR

Now that some general definitions for the system have been defined, more
detail as to the specific sensors and their applications will be given.
Although not an extrusion device, it is important to understand the miniature
batch-mixing sensor, which often serves as the starting point for formulation
development. Many materials that are utilized for pharmaceutical products
are essentially a combination of different material types, chosen to impart
the benefits of the individual properties. However, theoretical compounds
often do not perform as predicted because of material incompatibility,
unforeseen chemical reactions between components, or the unexpected
impartation of detrimental properties from individual components of the

blend. A torque rheometer with a batch-mixing sensor provides a means in which the user can simulate compounding on a small, tightly controlled system to offer a useful prediction of these combinations. In addition, the system provides a means in which the compounding conditions for known formulations can be optimized.

The standard mixer system consists primarily of a central bowl section hollowed out to resemble a "figure eight." Within each lobe of the "eight" a rotor blade is housed, the type of which is chosen depending upon the level of shear the user wishes to impart upon the sample. These blades are rotated in opposing directions and at varying speeds in order to promote good flow/ mixing and optimum shearing action between the two lobes. The mixer chamber is completed through the incorporation of a front and back plate, which essentially close off and seal the bowl at both ends. A loading device and ram is utilized to introduce the sample to the chamber and define the final volume within the system (the lower portion of the ram actually forms the top wall of the mixing chamber) (see Figure 6).

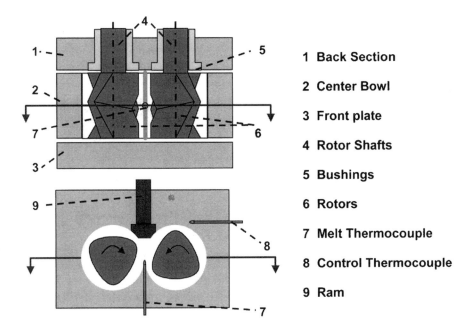

1 Back Section

2 Center Bowl

3 Front plate

4 Rotor Shafts

5 Bushings

6 Rotors

7 Melt Thermocouple

8 Control Thermocouple

9 Ram

Figure 6 Batch mixing system.

As has been described, the mixing system consists of three separate sections (back plate, center bowl, and front plate). In addition to allowing for ease of disassembly (useful in cleaning), this also allows for individual temperature control upon each of the three sections (a specific number of cartridge heaters are typically embedded within each section, along with a respective control thermocouple). This is important within miniaturized mixing systems as even slight temperature variations of a few degrees can cause dramatic changes in the results. This is important, as a significant degree of the heat energy resultant within the system is a result of the shear energy imparted from the motor and rotating screws (shear heating). For this reason, cooling is also available to attempt to maintain a proper temperature level (air-cooling channels are typically routed through the central bowl; air flow is automatically controlled via programmed solenoid actuation). And finally, as there will exist definite differences between the set temperature and the actual temperature of the material (mainly because of shear heating), a melt thermocouple is typically incorporated in the center bowl and protrudes within the chamber itself (refer to Figure 6). This signal provides an additional parameter that is utilized for material/process characterization.

There are a number of variables incorporated within the mixing system that allow for the manipulation of the resultant mixing mechanism. As noted, one that is extremely important would be the shear (or shear rate) induced upon the sample. This variable can be manipulated through the use of different rotor blades, the programming of different rotor speeds, and through bowl temperature modifications (as the viscosity of many materials is temperature dependent as well as shear dependent). Other variables that can affect the outcome (or efficiency) would include sample size or percentage loading, set temperature, and residence time during each stage of the experiment (the user may wish to incorporate a number of staged condition sets to expand the information obtained).

A number of testing procedures have become standardized in the usage of a torque rheometer/batch-mixing system combination. A few examples, along with the results obtained, are described below.

A.   Batch Differentiation

Pharmaceutical formulations can often be differentiated through the associated torque (relative viscosity) levels under varying degrees of shear. Typical usage for this type of test would be to predict the success/failure of a modification made to a "good" formulation (or standard). For example, material A with a certain level of "additive" has been processed within a torque rheometer/batch-mixer system and gives a respective rheograph (see Figure 7). In the interests of improving the compound's efficacy, the users are investigating the processing response of material A with different additive levels.

It is evident that specific changes can be noted between the torque curves for the two different formulations. The levels of torque for the two systems react differently within the first 4 min of testing, although after that point both compounds seem to stabilize at a certain similar value (approximately 23 N m). Beyond approximately 13 min, however, it is important to note that a significant increase in the formulation with limited stabilizer can be seen. On the other hand, the material with increased stabilizer continues on with a stable torque signal until almost 16 min. As a synopsis it could be concluded that by reducing the amount of additive within the formulation the users have reduced the amount of time before degradation will take place within the system by approximately 3 min. However, this does not mean that the experiment was a failure—it could be the case that the residence time for the actual manufacturing (extrusion) process can accept this reduction in "stable time."

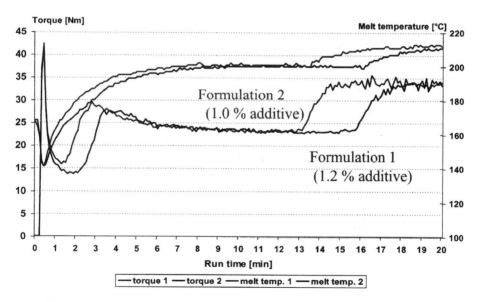

Figure 7 Batch differentiation rheograph.

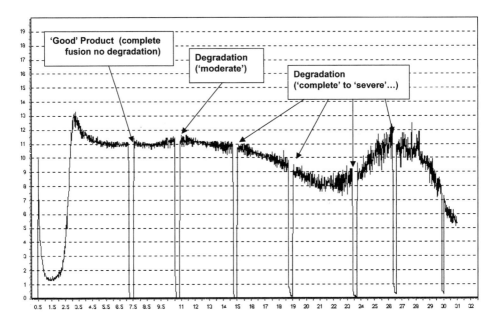

Figure 8 Sample evaluation with data collection.

B. Sample Compounding for External Evaluation

One important feature of a torque rheometer/batch-mixer system is the ability to compound small quantities of materials for external evaluation. A typical mixing system can operate with as little as 50 g of material, which is often sufficient for standard torque rheometry evaluation as well as sample collection during varying stages of the compounding process (see Figure 8). Specialized systems are also available that operate with as little as 6 g of material.

Within this test the standard procedure was modified to incorporate a collection of the material at specified time intervals for visual inspection.

C. Specialized Adaptations for Specific Testing

As has been mentioned, the torque rheometer system provides the user with great flexibility in testing protocol. This can prove exceptionally beneficial for very specific testing requirements in that the system can be readily adapted to meet these needs. One case in which a system has been modified that is used in the rubber industry is to determine the conductivity of rubber compounds as a measure for the dispersion of carbon black (1).

A common testing procedure within the rubber industry involves the blending of rubber with carbon black within a torque rheometer system. Based upon varying levels of shear a relative measure of the degree of an additive's incorporation can be made directly through the observation of the torque signal. This information gives a good indication of the dispersion of carbon black within the base material. Within this experiment, however, the batch mixer has been modified to incorporate a conductivity sensor in place of the common melt thermocouple (in the central bowl). The chamber wall then would act as a secondary electrode and, depending upon the dispersion of carbon black within the base resin, a specific value of conductivity will be measured. This method provides an additional manner in which the degree of dispersion for carbon black can be evaluated (see Figure 9).

III. TORQUE RHEOMETER WITH TWIN-SCREW EXTRUDER SENSOR

A torque rheometer with twin-screw extruder sensor is in many ways similar to the batch mixer combination. The system allows the user to blend various types of materials and additives for the production of a compound; however, an important difference is that this system allows simulation for a continuous production. In addition, more complex configurations and operations are available with the continuous system, which are not feasible for batch processes.

In general, twin-screw systems are utilized for continuous compounding (mixing) applications. This process may result in a final product being formed, or may act as an intermediary step in its development. Whatever the case, the addition of a torque rheometer system to the twin-screw extruder allows the user to simulate this task while maintaining a very tight level of control over the various parameters involved (screw speed, zone temperatures, feeding systems, postextrusion devices, etc.). In addition, experimental signals such as melt temperatures, pressures, and output rates are monitored to assist in an analysis of the development. Just as a batch mixer allows the user to investigate a material and simulated process, the twin-screw extruder allows for an actual small-scale compounding process to be performed and optimized in real-time under processing conditions and constraints.

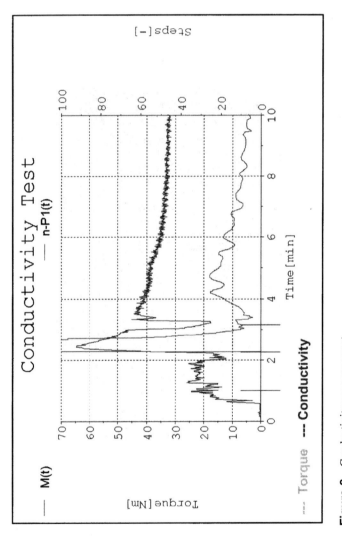

Figure 9 Conductivity measurement.

The following detail a few examples of common applications for twin-screw extrusion/torque rheometers systems:

A. Sample Production Within a Counterrotating, Conical Twin-Screw System

The importance of torque rheometer systems is evident, as has been applied for many years to PVC plastics used for medical devices, such as tubing. A good compliment to this test would be the utilization of a laboratory conical twin-screw system for the creation of a sample based on these PVC materials. Figure 10 shows how a correlation can be made between the residence time, or process time for the material within the extruder, as a complement to degradation information obtained from a batch-mixing system.

B. Clinical Trials Within a Corotating, Parallel Twin-Screw System

One of the most beneficial facets of the torque rheometer system is that it provides the user with a great degree of control over not only the processing parameters but also the environment within which the system is run. The units are often referred to as "lab-scale," clearly describing the fact that these

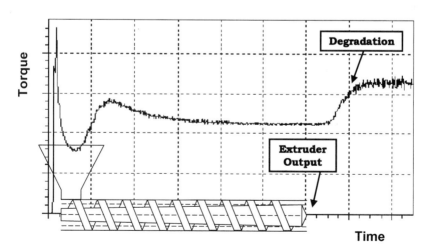

Figure 10 Residence time, degradation correlation for PVC compound from batch mixer applied to the extrusion process.

devices are smaller than production-sized counterparts. This reduction in size also means that in general all operations performed upon these systems can be accomplished with a greater degree of flexibility and ease—start-up, maintenance of the run, sample collection, shutdown, and cleaning. This can be of great interest for small-scale clinical studies, as commonly performed within the pharmaceutical industry.

In order to further facilitate this ability many systems have been designed to exaggerate this ease of usage. An example might be a corotating twin-screw extruder with 15- to 30-mm screw diameter with a clamshell-design barrel, which allows the users to fully disassemble and investigate all portions of the device. This provides great utility not only for the shutdown and cleaning of the unit but also in simple shear investigations—during a trial the system can be stopped (drive unit) and temperatures within the extruder lowered to solidify the materials within. The top portion of the barrel can then be removed and a visual examination of the sample's progression (and shear history) down the extruder screws can be made. Further morphology studies could then also be made through the collection and comparison of samples at varying positions within the extruder.

C. Microcompounding

A large portion of applications for torque rheometer systems stems from the research and development of new materials. One reason for this lies in the cost of these materials, which can be thousands of dollars per kilogram. An additional cause could be a lack of availability of large sample sizes, often the situation with new and experimental materials. Whatever the case, taking advantage of a system that utilizes samples in gram sizes as opposed to kilogram quantities is certainly beneficial. A microscale compounder does exactly this.

A microscale compounder is essentially a system that allows for the user to compound very small quantities of material in as close to a traditional extrusion style as possible (see Figure 11). Predominant systems currently on the market offer a conical extrusion design. Some provide a "backflow" channel, which enables the user to maintain the testing material within the system as long as desired, thus defining the residence time. The unit itself is typically instrumented to control and measure a number of different testing parameters (torque, temperature, and pressure), and visualization software is also available to assist with data handling and analysis. Studies are possible with as little as 6 g with this type of device.

An experiment performed using a microcompounder can be seen in Figure 12, which demonstrates a reactive extrusion process utilizing a

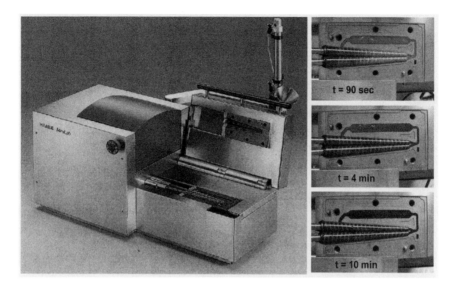

Figure 11   Twin-screw microcompounder with clamshell barrel and bypass valve.

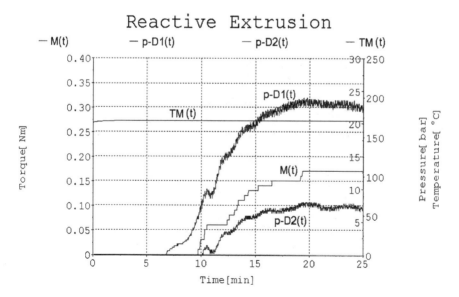

Figure 12   Reactive extrusion analysis—microcompounder.

bypass valve. The use of a bypass valve allows the material to recirculate in the system (2) and demonstrates that the time to reaction begins at approximately 7 min and stabilizes after approximately 18 min. This information can assist in assessing processing parameters for this material within production.

IV. TORQUE RHEOMETER WITH SINGLE-SCREW EXTRUDER SENSOR

The flow mechanisms for single-screw systems promote a constant, positive displacement of material, which enables a stable, surge-free pressurization to the exit (die) of the system and the creation of a "good," unvarying product. Applications can be quite varied for the single-screw extruder, as evidenced in other chapters of this book.

In addition to production simulation and quality control, a torque rheometer and single-screw extruder can also be utilized for a more absolute measurement of a sample's properties. In essence, absolute viscosity information can be obtained by assembling a system that simulates traditional piston-type capillary flow within an extrusion process. Both measurements utilize the same measuring principles; however, the mode in which the process melt is prepared and delivered to the capillary varies between the two (3).

In terms of the base measuring principle—a test sample is forced through a capillary (rod or slit) of known geometry with constant volumetric flow. The sample creates, depending upon its viscosity, a flow resistance that results in a pressure gradient along the capillary. This measurement of the pressure gradient provides a calculated shear stress. The volume flow of the material can then be measured and used to calculate the shear rate. The melt viscosity can then be calculated through division of the shear stress by shear rate. The calculations involved are dependent upon the geometries of the capillaries in question (different for slit and rod types), and in addition several corrections are necessary to account for error-inducing effects within the capillary (entrance pressurization) and to correlate for non-Newtonian flow.

Extruder-type capillary rheometers use a laboratory extruder to prepare the process melt stream and feed the capillary die. The variation of screw speed of the extruder sets different shearing rates. The flow rate of the test material is either measured using one of several balance methods (manual, semiautomatic, or automatic), or controlled through the use of a melt pump system. The advantages associated with using an extruder to simulate an extrusion process are inherently obvious.

V. SUMMARY

One of the most important features of any one piece of testing equipment is versatility. The ability to measure common characteristics of materials and processes is certainly essential; however, so is the adaptability of the equipment to adhere to changing needs of markets, materials, processes, and the ever-demanding user. Torque rheometer systems have, through the years, proven themselves over and over again in meeting these requirements for change and diversity, and are now being applied to pharmaceutical extrusion applications.

8
Process Design

Adam Dreiblatt
Extrusioneering International, Inc., Randolph, New Jersey, USA

I. INTRODUCTION

The term "melt extrusion" implies the processing of polymeric materials at temperatures above their glass transition temperature (T_g) in order to effect molecular-level mixing of active compounds and thermoplastic binders and/or polymers (1). These processes are increasingly being performed on twin-screw extruders as a result of their superior mixing capabilities and narrow residence time distribution (2,3). A narrow residence time distribution is critical when processing thermally sensitive active compounds.

The versatility of the twin-screw extruder is exemplified by the wide range of industries and applications in which it is found. The same basic components are used in the polymer processing industry to compound hydrocarbon-based polymers and fillers; in the food processing industry for extrusion cooking of starch-based biopolymers; and in the confectionery industry for continuous cooking of polysaccharides. In each case, the same fundamental unit operations are executed within the confines of the extruder barrel:

Feeding raw materials
Melting/plasticization
Mixing/dispersing
Venting
Die pressurization

These same unit operations are required for application of twin-screw extruders in the pharmaceutical industry. The specific details of each unit operation as it applies to pharmaceutical products will be discussed in this chapter.

II. FEEDING RAW MATERIALS

Raw materials (active compound, binder/polymer, excipients, and additives) must be introduced into the extruder; thus, there are some requirements for feeding. The type of extruder, single-screw or twin-screw, will determine the feeding requirements.

A. Single-Screw Extruders

Single-screw extruders are referred to as drag flow pumps (4). The drag flow conveying mechanism depends on a frictional relationship between the raw materials and the extruder surfaces: The materials must adhere to the barrel wall and must not adhere to the screw. Thus, the conveying characteristics of a single-screw extruder are directly related to raw material properties: particle size, particle size distribution, particle shape, particle–particle interactions, etc. This requires all raw materials in a blend (for example, binder/polymer, excipients, and active compound) to be similar in particle size and density. Thus, raw materials that are sticky or slippery will not feed well in a single-screw extruder.

B. Twin-Screw Extruders

Corotating twin-screw extruders are also considered drag flow pumps; however, the conveying mechanism does not depend on a frictional relationship between the raw materials and the extruder surfaces (5). Counterrotating twin-screw extruders are not considered drag flow pumps, but rather positive displacement pumps. In both of these cases, the extruders are insensitive to raw material properties as far as feeding is concerned. Raw materials can be sticky, slippery, and have dramatically different particle sizes and densities.

Flood Feeding

Most single-screw extruders are flood fed; that is, the raw materials are placed in a hopper situated over the screw such that the hopper is filled with the raw material (4). As the screw rotates, the volume of material conveyed is directly related to the screw rotational speed (revolutions per minute), the pitch, and free volume of the screw. As the screw speed is increased, so is the volume of material displaced. In this case, the screw speed determines the output of the extruder as long as the screw is kept full in the feed area. The feeding behavior of a single-screw extruder is analogous to the feeding of a tablet press.

Feeding raw materials into a single-screw extruder places significant restrictions on the raw materials themselves, whereas the extruder requires only that the screw be maintained full in the feed area. If the screw speed and raw material characteristics can be maintained constant, the output of single-screw extruders is relatively consistent with respect to time. This is a critical requirement, e.g., in the production of multiparticulates where constant extruder output is necessary to produce constant extrudate diameter and exit velocity from the extruder die. Variations in raw material density, particle size, or other property (which would affect the frictional relationship between raw material and extruder surfaces) manifest themselves in extrusion instabilities, often referred to as "surging."

Starve Feeding

Twin-screw extruders do not depend on the frictional relationship of raw materials for conveying; these extruders do not need to be flood fed. They are designed to be starve fed, where the extruder screws are not filled in the feed area (4). In starve-fed extruders, a separate metering device is used to control the rate of raw materials delivered to the extruder per unit time.

This concept allows the extruder screw speed to operate independently of output; alternately, the extruder can operate at various outputs for a given screw speed. This additional degree of freedom provides enormous process capabilities; however, the extruder output now depends on the accuracy and precision of these feeding device(s).

Feeding Equipment: Volumetric vs. Gravimetric

Feeding equipment for starve-fed extruders can be volumetric or gravimetric. It is not within the scope of this chapter to explore the technology of feeding equipment (6). It is, however, within the scope of this chapter to discuss how feeding equipment directly influences the performance of the extrusion process.

Most feed materials in pharmaceutical applications are in powder form as opposed to the plastics processing industry where most feed materials are in a pellet form. There are several feeder technologies available for metering powders: screw feeders, vibratory feeders, and belt feeders. Of these, screw feeders are the most common and are available either as volumetric devices or as gravimetric devices.

Volumetric feeders operate at a constant screw speed to deliver a constant volume per unit time. As such, the mass flow will vary as a function of differences in material density. This may occur within the batch itself and/or during the refilling of the feeder. Volumetric feeders generally provide accuracy and precision within $\pm 2\%$ of setpoint. Gravimetric feeders utilize the same volumetric screw feeder mounted on a load cell or scale. A microprocessor-based controller then monitors the loss in weight per unit time from the scale or load cell, and adjusts the speed of the screw to achieve the target mass flow setpoint. These systems generally provide accuracy and precision within $\pm 0.5\%$ of setpoint.

If all raw materials are blended together in a batch operation and introduced into the extruder using a single feeder, the feeder variability cannot affect the composition exiting the extruder. Neglecting the segregation of components within the feeder itself, the content uniformity of the product entering and exiting the extruder is determined by the blending operation, not by the extruder or the feeding equipment. Variability in feed rate, in this case, produces unstable output with respect to time, because the extruder is starve fed. As mentioned above, consistent extruder output is a critical requirement for producing multiparticulates.

If, on the other hand, multiple feeders are utilized to separately meter components into the extruder as is commonly practiced in the polymer and food industries, the content uniformity of the product exiting the extruder is directly affected by the accuracy and precision of the individual feeders (7). For example, if each of two feeders has an accuracy and precision of $\pm 0.5\%$ (relative standard deviation value typically specified by feeder manufacturers at two standard deviations), then the content uniformity entering the extruder can vary with respect to time by this amount. While well within USP guidelines, this concept represents more a cultural obstacle rather than a technical one for an industry based on batch weighing.

Short-term variations can be eliminated via backmixing within the extruder screws; however, the consequence of this type of screw design implies a wide residence time distribution. Long-term feeder variations should be avoided at all costs, as these cannot be dampened by the extruder if the variations are in excess of the mean residence time within the extruder (8).

The concept of separately metering components offers significant advantages. For example, a thermally sensitive active compound can be introduced to the last part of the extruder where it is mixed with the already molten polymer. The residence time of the active is substantially reduced as a result of separate metering of components. Other advantages can be realized by eliminating the blending operation, using gravimetric feeders to formulate materials as they enter the extruder. It is then up to the extruder to produce a homogenous blend.

Liquids can also be introduced into the extruder for mixing with solids or injected downstream directly into the molten polymer(s). Metering of liquids requires the same degree of accuracy and precision as was mentioned for feeding solids. Positive displacement pumps are preferred; they can be equipped with mass flow meters for monitoring and control or supplied as a loss-in-weight system using a scale-mounted reservoir.

III.   MELTING/PLASTIFICATION

The key feature of all screw-type extruders is their ability to efficiently melt and plasticize materials. As an energy conversion device, screw extruders use two forms of energy transfer: thermal and mechanical.

A.   Conduction Melting with Forced Melt Removal

This melting mechanism is directly influenced by thermal energy input where a polymer film is generated on the inner barrel surface as a result of heat conduction through the barrel (energy is supplied from either electrical resistance heaters or circulation of thermal fluids). The rotation of the screw flight(s) displaces the melt film, mixing the molten material into the bulk. This process depends on the thermodynamic properties of raw materials, particle size (and shape), percent fill of the extruder screw, and thermal gradient between the material and barrel surface. The thermal contribution to melting can be quantified on a unit mass basis:

Specific thermal energy (kWh/kg)
     = Thermal energy input (kWh/hr) ÷ Mass flow (kg/hr)

The surface/volume ratio for thermal heat transfer in screw-type extruders is several orders of magnitude greater than in conventional batch-type equipment (9).

B.   Dissipative Mix Melting

Screw-type extruders are designed to convert electrical energy from the drive motor into mechanical energy via viscous dissipation. Melting occurs as solids are confined to decreased clearances (twin-screw extruders) and/or decreased free volume (single-screw extruders). In corotating twin-screw extruders, staggered kneading discs are employed to melt materials (10,11). Melting typically occurs rapidly within several kneading discs in a twin-screw extruder and over several diameters in the compression section of single-screw extruders. This melting mechanism is directly influenced by mechanical energy input from the drive motor and can be quantified on a unit mass basis:

Specific mechanical energy (kWh/kg)
= Consumed motor power (kWh/hr) ÷ Mass flow (kg/hr)

The thermodynamic properties of the raw materials, raw material temperature, and particle size are all key parameters contributing to the efficiency of the extruder in melting via viscous dissipation. Lubricating additives and materials with low melt viscosity can adversely affect the introduction of mechanical energy and thus the melting process.

In practice, very few materials are melted using either one or the other mechanism. Most materials require both thermal and mechanical contributions to achieve a complete melt. Values for specific mechanical energy input range between 0.1 and 0.4 kWh/kg for most thermoplastic materials. Values for specific thermal energy input are approximately an order of magnitude less than specific mechanical energy input.

When a blend of two materials with dramatically different melting points is processed, the material that melts first can potentially experience some degradation by the time the second component has melted (e.g., active compound with low melting point and polymer with high melting point). In this case, it would be beneficial for the material with a lower melting point to be introduced downstream into the extruder through a secondary feed opening, into the already molten material. This second component would melt rapidly at the higher temperature and would also have less residence time.

Polymers that melt at relatively low temperatures (i.e., waxes) may prematurely melt in the feed opening of the extruder barrel, causing problems with feeding. In these cases, the feed barrel is kept cool with increasing temperature in the subsequent barrel modules.

IV. MIXING

A. Dispersive Mixing

Screw-type extruders produce a fluid shear stress as a result of the velocity gradient of rotating screw(s) and a stationary barrel. It is this shear stress that is responsible for reducing the domain size of the minor phase (e.g., active compound), also referred to as dispersion or dispersive mixing (12). In a simplified form, the magnitude of the applied stress is a function of the shear rate and the melt viscosity:

Shear stress (kPa) = Shear rate (sec^{-1}) × Viscosity (Pa sec)

From this relationship, it is apparent that it becomes more difficult to achieve high shear stress with low-melt-viscosity materials than with high-viscosity materials.

Shear rate, describing the velocity gradient between two surfaces moving at different speeds (e.g., screw and barrel wall), is a function of screw outside diameter, screw speed, and gap:

Shear rate (sec^{-1})
$$= \frac{[\pi \times \text{Screw diameter (mm)} \times \text{Screw speed (rpm)}]}{[\text{Gap (mm)} \times 60 \text{ (sec/min)}]}$$

where the gap is defined as the distance between the moving surfaces (e.g., the mechanical clearance between screw and barrel or between screw/screw in twin-screw extruders).

The range of shear rate achievable in screw extruders is between 10 and 10,000 sec^{-1}. High values are achieved with large screw diameter, high screw speed, or small gap. Thus, for a specific extruder diameter operating at a specific screw speed, the highest shear stress occurs where the viscosity is the highest, i.e., during the initial melting of polymer(s). This concept provides the motivation to feed all components into the primary feed opening of the extruder to provide the highest degree of dispersive mixing. Conversely, heat-and/or shear-sensitive compounds can be fed downstream in the extruder, thus avoiding the high shear stresses experienced during the melting process. When active compounds are fed downstream into the already molten polymer, the maximum possible shear stress is much lower because of the lower melt viscosity.

The continued application of shear stress within the extruder screw produces a subsequent decrease in minor phase size, also referred to as morphology development. The final structure is "frozen" in place when the polymer(s) are cooled below their glass transition temperature (T_g) after being discharged from the extruder. The resulting morphology is responsible for producing a particular dissolution profile or bioavailability.

The extrusion parameters that influence shear stress are used to control the morphology (i.e., dispersive mixing). These parameters are listed in order of significance:

1. Screw design. The configuration of the screw (including overall *L/D* and mechanical clearances) will determine how much of the

material will experience high shear stress. Flow-restriction devices are used in both single-screw and twin-screw extruders to force material through small clearances. As mentioned previously, the modularity of twin-screw extruders provides an additional degree of freedom to modify the shear intensity of the screw by installing multiple flow restrictions.

2. Screw speed. Directly proportional to shear rate. This parameter produces a linear effect when the extruder is flood fed (e.g., single-screw extruders) and a nonlinear effect when the extruder is starve fed (because increasing screw speed at a constant feed rate will also produce a decrease in the filled screw volume).

3. Feed rate. Directly proportional to shear rate for flood-fed extruders, because the only way to increase feed rate is to increase screw speed. The effect of increased feed rate on starve-fed extruders is to decrease the "effective" shear rate, because the filled screw volume will increase at a constant screw speed. The increase in filled screw volume means that more material is occupying that portion of the screw channel with larger gap (i.e., in the center of the screw channel), as opposed to when the screw is nearly empty when most of the material is occupying that portion of the screw channel with smaller gaps (i.e., clearance between screw and barrel).

4. Barrel temperature. Increasing barrel temperatures will decrease melt viscosity, thereby decreasing the shear stress.

B. Distributive Mixing

Distributive mixing can also be viewed as homogenization, where the concentration of the minor phase (e.g., active compound) is constant throughout the volume. Distributive mixing is responsible for the content uniformity of a particular constituent, whereas dispersive mixing is responsible for the size and size distribution of a constituent (13). Distributive mixing is achieved in an extruder by the interchange of discrete volumes contained within the extruder screw. Twin-screw extruders contain multiple screw channels (i.e., discrete volumes), which can be split and recombined. Screw elements are available for twin-screw extruders, which promote distributive mixing by modifying the number of discrete volumes; single-screw extruders contain only one screw channel and therefore rely on interruptions in the flow channel to provide reorientation.

V. VENTING

The removal of volatiles under vacuum is a common unit operation for most plastics compounding applications. The volatiles tend to be residual monomers, sizing agents used for reinforcing fillers, residual moisture, etc. Venting efficiency depends on melt temperature, which will influence the vapor pressure of the volatile(s); vacuum level; and residence time under vacuum. Twin-screw extruders can be operated with multiple vents operating at different vacuum levels to increase residence time under vacuum.

Depending on the processing temperature, residual moisture can/will be evolved as well as entrapped air. If the temperature is high enough to cause the degradation of active and/or polymer to occur, these degradation products can also be volatilized.

The screw(s) in the vent area needs to operate with the lowest degree of fill to maximize the surface/volume ratio for devolatilization efficiency. In order to sustain a vacuum, the screw must be 100% filled both upstream and downstream of the vent. Restrictive screw elements are used upstream of the vent to create a "melt seal" (section of screw that is filled 100%); the die usually becomes the vacuum seal downstream. The maximum sustainable vacuum then becomes a function of the pressure drop across the melt seal. Vacuum levels less than 100 mbar absolute are easily achieved.

VI. DIE PRESSURIZATION

The majority of melt extrusion applications involves the shaping of the molten material through an orifice or multiple orifices. The end-use application will dictate what die geometry will be used:

1. Transdermal applications will use a sheet die to shape the molten material as it is deposited onto chilled rolls. The final sheet thickness and width are product specific.
2. Products that will be milled for tablet compression will typically be extruded as continuous strands. The strands provide increased surface/volume ratio for cooling and solidification as well as provide a means for product handling for feeding to a mill. In this case, the strand diameter and the number of strands are not critical, as the milling operation will determine the particle size for compression. Typical strand diameter is 3 mm out of convenience, as this is the standard strand diameter used for most thermoplastics.

3. Multiparticulates are extruded as continuous strands and pelletized in-line. In this case, the strand diameter is critical to provide a surface/volume ratio for enhanced dissolution. Strand diameters are typically 1 mm or less; the number of strands is often determined by the configuration of the in-line pelletizer.

The extruder must generate the pressure required to overcome the resistance of the die, a function of die open area and land length (defined as the length of die that is at final cross section). The pressure drop is then determined by the melt viscosity and flow rate. Pressure drop through extrusion dies can be calculated for simple shapes (e.g., round holes) if the rheological behavior of the melt is known. Transitions from the extruder screw(s) to the final die should be streamlined with minimum free volume to avoid stagnation points where melt can degrade. If a large die open area is required (e.g., large number of strands or to produce a wide sheet), the design of the die becomes complex to produce uniform velocity across the die width.

Whereas the mechanical limit of extruders for continuous operation is in the range of 200 to 350 bar, die pressure requirements should be minimized whenever possible (e.g., largest die hole diameter, shortest land length, largest number of die holes, etc.). One of the consequences of die pressurization is the accompanying melt temperature rise from viscous dissipation. Fully inter-meshing and counterrotating twin-screw extruders are the most efficient in generating pressure with minimum temperature rise, followed by single-screw extruders and fully intermeshing and corotating twin-screw extruders. Melt pumps can also be installed at the end of the extruder for those applications requiring either very high die pressure (e.g., very small orifices) or precise metering (e.g., direct extrusion of sheet, film, etc.). Gear pumps or single-screw extruders can be coupled with corotating twin-screw extruders to minimize the temperature rise and/or provide consistent gauge control.

VII. PROCESS LENGTH

Given the screw outside diameter and channel depth (or inside diameter), the free volume of the extruder is then determined by of the length of the screw. Extruder length is referred to in dimensionless units as a function of screw diameter:

Extruder length (L/D)
$=$ Absolute length(mm) $\div$ Screw diameter (mm)

The extruder length required for a particular application will depend on how many different unit operations are needed. Because twin-screw extruders are modular in design, a sequence of unit operations can be assembled for each application. Twin-screw extruder barrel modules are manufactured in lengths approximately equal to four times the screw diameter (4D). Each unit operation can be assigned to a single barrel module and several barrel modules can be dedicated to one unit operation. Assumptions must be made for backup length (length of screw required to generate pressure), melting and mixing. It is a safe assumption to assign two modules to a unit operation if in doubt. Some examples:

1. A simple process can be envisioned where all materials are premixed and fed into the main feed port of the extruder (4D). The polymer component is melted (4D), an active compound (particulate component) is dispersively mixed (4D), and the resultant homogenous molten mass is pressurized through a die orifice (4D). The minimum process length for this application is 16D.

2. A complex process would require feeding of polymer(s) only in the primary feed port (4D) where they are subsequently melted (4D). Because of the low melting point of a particular polymer, an additional barrel module is installed as a thermal barrier to the feed zone (4D). Active compound(s) are then fed into the molten polymer (4D) where they are intimately dispersed (4D). An additional barrel module is assigned to mixing as a result of the low melt viscosity of the polymer (4D). The materials are then subject to vacuum (4D) to remove entrained air and residual moisture. Two barrel modules are dedicated to die pressurization because of the low melt viscosity and small die orifices used for extruding multiparticulates (8D). The minimum process length for this application is 36D.

There is a limit to the maximum length of a twin-screw extruder, determined by the torque limitation of the screw shaft and/or maintenance of clearances at extreme L/D. This limit is between 50 and 60 times the screw diameter for intermeshing twin-screw extruders and therefore limits the maximum number of unit operations (or limits the number of barrel modules available for a particular unit operation) possible within the extruder. It is also possible to operate two or more extruders in series to overcome this limitation.

VIII. EXTRUDER CAPACITY

As the primary dimension for specifying an extruder, the screw diameter determines the capacity range for both single- and twin-screw extruders. The factors that influence the production capacity of an extruder in a pharmaceutical environment are not as straightforward as in other industries because of the batch-oriented processes upstream (weighing and blending) and downstream (tableting/coating, encapsulation, etc.).

Whereas the batch size for a given product is determined by the volume of blending equipment, the production rate of the extruder becomes a variable. The batch can be run at a high production rate (i.e., large-diameter extruder) for a short period of time or it can be run at a relatively low production rate (i.e., small-diameter extruder) over an extended period of time. The time to process the batch together with the batch size will be the determining factors in sizing an extruder. Additional factors to consider are capital equipment costs (favoring small extruder and longer run time) and labor costs (favoring larger extruder with shorter run time). Extruders are manufactured for production rates as low as 1 kg/hr and as high as 20 tonnes/hr.

IX. PROCESS MODEL

Whether using a single-screw or twin-screw extruder, for a given formulation the interaction of extruder process variables produces energy transfer to the raw materials over a finite residence time with a resulting increase in product temperature. In turn, the specific energy input (both mechanical and thermal), residence time, and product temperature produce a morphology of the disperse phase (e.g., active compound) and/or crystallinity. It is this resulting morphology that then produces a particular dissolution behavior and/or bioavailability.

Attempts have been made to correlate product attributes (e.g., dissolution profile) directly with extruder variables such as screw speed, feed rate, and barrel temperature, without success. This is because increased extruder screw speed, e.g., does not directly influence the dissolution, but rather increases the specific mechanical energy input. It is the increased specific mechanical energy input that produces a decrease in minor phase size (and subsequent increase in surface area). And finally, it is this difference in morphology that is responsible for the resulting increase in dissolution rate.

Response surface methodology (RSM) is a valuable method to "map" the influence of each part of the process. Van Lengerich (14) proposed such a model for extrusion of food starches whereby:

> Final product properties ("target parameters") can be directly correlated with physical and/or chemical transformations ("structure").

Physical and/or chemical transformations ("structure") can be directly correlated with the introduction of specific mechanical and thermal energy, residence time, and product temperature ("system parameters").

The introduction of specific energy ("system parameters") can be directly correlated with extrusion parameters ("process parameters").

Figure 1 illustrates this concept using the process parameters available on a twin-screw extruder and pharmacological product attributes for melt extrusion applications.

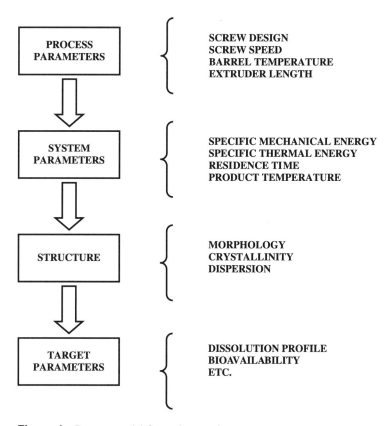

Figure 1 Process model for melt extrusion.

X. PROCESS TROUBLESHOOTING

Process disturbances that adversely affect the melt extrusion process can be divided into two categories: consistency and reproducibility. Consistency refers to disturbances that can occur within one batch during a continuous operation whereas reproducibility occurs from batch to batch. These concepts are not specific to melt extrusion; however, they form a useful framework to distinguish between assignable causes of process disturbances. In most cases, these "upsets" to the system can be attributed to mechanical failures, which can also occur with any piece of process equipment. The most common failures will be discussed as well as potential preventive measures.

A. Consistency

Disturbances can occur during a continuous melt extrusion process as a result of mechanical failures or operational disturbances:

1. Mechanical failures in extruder heating/cooling systems
   a. Extruder barrel cooling solenoid valves that do not close completely as a result of sediments in the cooling water. This type of failure results in extruder barrel temperature deviation beyond acceptance limits. Preventive measures involve the installation of filters in the coolant supply line and monitoring of coolant quality.
   b. Extruder barrel and/or die heaters that burn out result in temperature deviation beyond acceptance limits. Preventive measures involve the monitoring of condition of heating elements.
2. Disturbances in feed rate
   a. Feed rate can vary during the refilling of the feeder; excessive disturbances to the feeder during the refill process, and/or excessive length of time for completion of the refill can produce potentially large variations in feed rate. Preventive measures require rapid and reproducible refill systems as well as increasing the refill frequency.
   b. Differences in raw materials within a single batch (particle size, density, etc.) will cause variations in feed rate from a volumetric feeder. Preventive measures require consistency of raw materials throughout the batch.
   c. Accumulation of powder anywhere between the feeder and the extruder (e.g., in the extruder feed hopper) can be dislodged and

fall into the extruder screw producing an instantaneous high feed rate as well as adversely affecting content uniformity. Preventive measures require monitoring of powder flow from the feeder to the extruder to avoid any possible accumulation.

B. Reproducibility

Disturbances that can cause batch-to-batch variation:

1. Mechanical wear of components (extruder screws, barrel, die) can affect the introduction of specific energy from increased mechanical clearances. Preventive measures require monitoring and documentation of equipment condition.
2. Differences in raw material characteristics from batch to batch (particle size, bulk density, etc.) will result in differences in mass flow when using a volumetric feeder. Preventive measures require consistency of raw materials from batch to batch.
3. Differences in polymer rheology from different lots of polymer must be characterized during the product development phase and documented during validation.
4. Incorrect screw assembly for modular twin-screw extruders is possible, because there are no absolute methods to verify screw-assembly sequence. Preventive measures include operator training and possible marking of elements with assembly sequence.

XI. CONCLUSION

A melt extrusion process can be assembled as a sequence of unit operations when using twin-screw extruders because of their positive conveying characteristics. The first decision to be made is whether to feed the extrusion process from a blending operation or to use separately metered ingredients. Although commonplace outside of the pharmaceutical industry, separate metering of components into a continuous extrusion process presents unique challenges for a batch-oriented culture.

Once inside the extruder, the polymeric components and/or active compound(s) are melted via a combination of thermal and mechanical energy within the confines of the extruder screw. Molecular-level mixing can be achieved to produce solid dispersions or controlled-release matrices; the degree of mixing is a function of the screw design and operating parameters.

Modular barrel design together with segmented screws provides unlimited combinations for tailoring the mixing capabilities of the equipment to the requirements of the product. Because the extruder length is also variable, the free volume and resulting residence time for mixing can be optimized for each product.

The homogenous molten mass can then be exposed to vacuum to remove volatiles or entrapped air prior to die extrusion. Finally, the material is shaped into strands, sheet, or other shapes for subsequent processing (e.g., continuous strands can be cut into lengths for multiparticulates or milled for tablet compression).

The small free volume and precise control of energy and mass transfer available in an extruder offers the pharmaceutical industry an interesting alternative to high-shear granulating equipment. The narrow residence time distribution of a melt extrusion process is not possible to reproduce on conventional batch-processing equipment. Active compound(s) can be processed at very high temperatures in a twin-screw extrusion process because the melt residence time is short (typically less than 1 min) and reproducible from the self-wiping action of the screws. These unique features provide formulation opportunities for the development of controlled-release products and establishing melt extrusion as a viable drug delivery platform.

REFERENCES

1. McGinity JW, Zhang F, Repka MA, Koleng JJ. Hot-melt extrusion as a pharmaceutical process. Am Pharm Rev 2001; 4:2, 25–36.
2. United States Patent No. 5,456,923. Nippon Shinyaku Company, Ltd., Japan.
3. Grünhagen HH, Müller O. Melt Extrusion Technology. Pharm Manuf Int 1995:165–170.
4. Wildi RH, Maier C. Understanding Compounding. Munich: Carl Hanser Verlag, 1998.
5. White J. Intermeshing co-rotating twin-screw extrusion. In: White J, ed. Twin Screw Extrusion. Munich: Carl Hanser Verlag, 1991:195–270.
6. Wilson D. Feeding Technology for Plastics Processing. Munich: Carl Hanser Verlag, 1998.
7. Maloncy J. Weigh-feeders and quality assurance. In: Setpoint, Ktron Technical Bulletin, October 1994.
8. Curry J, Kiani A, Dreiblatt A. Feed variance limitations for co-rotating intermeshing twin-screw extruders. Int Polym Process 1991; 6:2, 148–155.
9. Davis WM. Heat Transfer in Extruder Reactors. In: Xanthos M, ed. Reactive Extrusion. Munich: Carl Hanser Verlag, 1992:257–282.

10. Andersen PG. Mixing practices in corotating twin-screw extruders. In: Manas-Zloczower I, Tadmor Z, ed. Mixing and Compounding of Polymers. Munich: Carl Hanser Verlag, 1994:679–705.
11. Dreiblatt A, Eise K. Intermeshing corotating twin-screw extruders. In: Rauwendaal C, ed. Mixing in Polymer Processing. New York: Marcel Dekker, 1991:241–266.
12. Manas-Zloczower I. Dispersive mixing of solid additives. In: Manas-Zloczower I, Tadmor Z, ed. Mixing and Compounding of Polymers. Munich: Carl Hanser Verlag, 1994:55–83.
13. Agassant JF, Poitou A. A kinematic approach to distributive mixing. In: Manas-Zloczower I, Tadmor Z, ed. Mixing and Compounding of Polymers. Munich: Carl Hanser Verlag, 1994:27–54.
14. Van Lengerich B. Entwicklung and Anwendung eines rechnerunterstuetzten systemanalytischen Modells zur Extrusion von Staerken and Staerkehaltigen Rohstoffen. Ph.D. dissertation, Technical University of Berlin, 1984.

9
Melt Pelletization

Christopher C. Case
Conair Reduction Engineering, Pittsburgh, Pennsylvania, USA

I. INTRODUCTION

Pelletization is a downstream operation for a melt extrusion process where materials are taken from a device such as an extruder or melt pump, pumped through a die, cooled, and formed into a "pellet." Many products enter their first solid or densified state in a "pellet" form. This pellet, typically 3 mm in size or less, is then processed in another extruder, or molding machine. The main goal of pelletization is to facilitate the consistent feeding, transport, and packaging of the pellets. Generally, the pellet itself is made up of multiple components that have been compounded in a device, such as a twin-screw extruder, so that the material has desired properties.

Primarily used in the plastic industry, pelletization came into being due to a few simple facts. Many products needed to be transported from one place to another in a uniform particle size. Also, the pellet may need to be subsequently fed into another piece of equipment for further processing. Finally, the particle size often needs to be very similar to achieve an accurate and continuous feed rate. All factors that make a pellet desirable for a plastic pellet also apply to a pharmaceutical process.

Nearly every machine built for the pelletizing process for pharmaceutical applications is constructed of stainless steel product contact areas. Out of necessity, however, some components such as pneumatic cylinders, bearings, and seals might still be manufactured from oxidizing materials.

Pelletizers come in several designs and are selected based upon the material characteristics, throughput rate, and lot size. To more clearly describe the types of pelletizers and the related process operation, the following is a simple description of each commercially available type.

II. STRAND PELLETIZERS

The strand pelletizer was developed in the 1950s to cut strands into a canister-like final product. This is accomplished by forcing the product through a strand die that produces a "spaghetti"-like strand(s). The strand is then pulled through some type of a cooling medium by the feedrolls of the pelletizer and cut to length. The diameter of the final pellet is controlled by the diameter of the orifice and the speed at which the feedrolls are pulling the strand. Almost all products will be "drawn down" in diameter upon exiting the die. The feedroll speed is usually set in a ratio to the rotor or cutting head based on the number of cutting edges and the desired pellet length (see Figure 1).

Prior to the pelletizer, the strands must be cooled. For the plastics industry, the strands are typically drawn through a water bath by the pelletizer feedrolls. Upon exit of the water bath and before the pelletizer, an air stripper device removes moisture from the surface of the strands by means of either compressed air, an air blower, or by vacuum.

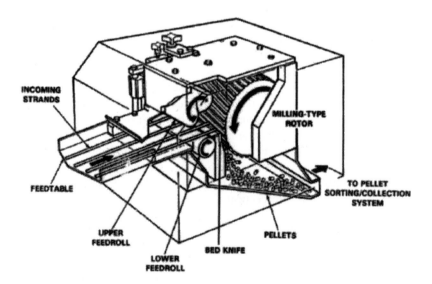

Figure 1   Closeup strand pelletizer.

As many pharmaceutical products cannot contact water, the cooling bath is often replaced with a cooling belt or static cooling table. Cooling belts can be stainless steel with liquid "spray" cooling underneath the belt to facilitate additional heat exchange. Belts can also be constructed of a Food and Drug Administration (FDA)-approved covering, or plastic mesh. Auxiliary blowers utilizing ambient or refrigerated air streams can specified to assist the cooling process (see Figure 2).

If the strands exhibit enough elasticity, another alternative is to utilize a series of grooved stainless steel rolls that are cored for liquid cooling. The "grooving" of the rolls will increase the surface contact for better heat transfer capabilities, as well as assist in sizing the shape of the strands.

Waterslide pelletizing is a system using a modified strand pelletizer cutting chamber coupled with an inclined trough where water flows from the end nearest the die head down toward the pelletizer at a slight downward angle.

The "Wet Cut" version diverts the water into the discharge chute just prior to pelletization and then reintroduces the pelletized product into the water for postcut cooling. The pellet/water slurry is then fed into a dewatering/ drying system to separate the water from the pellets. Some systems utilize a

Figure 2 Strand pelletizing system with belt conveyor.

forced hot air system to remove moisture from the pellets while others rely upon the residual heat left in the polymer to flash off the remaining moisture.

In contrast, the "Dry Cut" system removes the water from the strands in the waterslide and then strips the residual water off the strands using an integral vacuum air knife. The product is pelletized and discharged to a conveying system. The designation of "Dry Cut" is due to the fact that the pellets are not introduced back into water after pelletization.

III. DIE FACE PELLETIZERS

An alternative to strand pelletization is a process referred to as "die face" pelletizing. With this method, the strand is cut at the die face in a molten form and cooled via air or liquid. The advantage of this system, if workable, is that the problem of strand breakage during the cooling phase is eliminated. Die face pelletizing is not as wide spectrum with regard to the range of materials that can be processed, and start-up procedures need to be defined more carefully. However, if the process is amenable, die face pelletizing is often preferred over the strand systems defined above.

A. Air Quench Die Face Pelletizers

A die face pelletizing method that is often preferred for pharmaceutical applications is where the melt stream exits horizontally through a streamlined die and is cut by a flywheel or spring loaded cuter at the die face. After the "cut," pellets are pneumatically conveyed to either a vibratory cooler/ separator or fluidized vibratory unit, where the cooling process is finished. Additional product drying is possible if a heated air supply is supplied. Air quench pelletizing is ideal for formulations that have a high level of inert materials, such as dessicant compound. Air quench pelletizing should be tested for the specific products because many formulations cannot be processed this way, as many products tend to smear at the die face. This type system works well for higher viscosity and highly filled process melts (see Figure 3).

For some materials that have a tendency to smear, the use of Vortec™ tubes to direct chilled air at the die face facilitates die face cutting. Another alternative is to utilize atomizers to mist the die area for borderline formulations. Pellets produced via the air quench method are often slightly deformed in cooling, and are generally deemed not as aesthetically appealing as other methods described.

Figure 3 Air quench pelletizing system in plastics format.

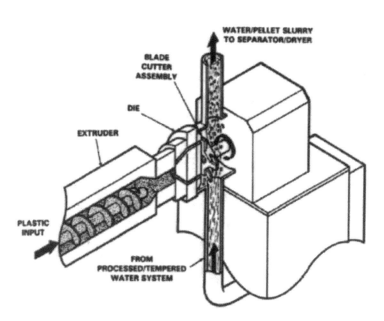

Figure 4 Underwater pelletizer schematic.

B. Underwater Pelletizers

The underwater pelletizing method is exactly what the name describes. The molten product is fed through a die with a series of holes in a circular pattern. When the product emerges from the die, it is cut into pellets by the rotating blades, with the pellet immediately solidified in the water. The water/pellet slurry is then pumped to a spin dryer for dewatering and surface drying, returning the process water to central system. This system produces a very unique pellet in that it is spherical in shape, resulting from the product being cut in a submerged environment. The underwater pelletizer is recommended for lower-viscosity materials that can withstand water contact and cannot be cut in air (see Figures 4 and 5).

Any product that retains temperature is a candidate for underwater pelletization. However, products that have a quick quench time, or processes that have intermittent flow generation are not suited because of the potential of die hole freezeoff. The direct contact of water with the die causes the potential for die hole "freezing" when the flow out of one of the holes is interrupted. The problem is particularly acute during start-up. For

Figure 5 Underwater pelletizing system.

this reason, the start-up sequence of the extruder, cutter, and water flow is critical and generally requires a PLC for consistent and repeatable process management.

C. Water Ring Pelletizers

Another die face design is a water ring pelletizer, where the pellets are cut at the die face in air and "flung" into a slurry discharge, which is pumped into a centrifugal dryer where the pellets are separated from the water, similar to the underwater unit described above. Most commercial water ring pelletizers discharge vertically downward, which requires that the melt stream be pumped through a 90° angle adapter. This design is generally not suited for pharmaceutical applications, as the increased residence time and the flow stagnation often adversely affect the product properties, and make cleanout more difficult. There are, however, commercial units that integrate a horizontal die design similar to the underwater pelletizing method described above, which eliminates this problem (see Figure 6).

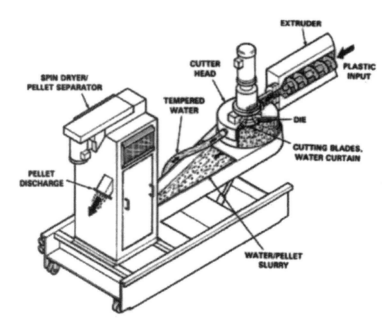

Figure 6 Water ring pelletizing system.

IV. VIBRATING RING DROPPO PELLETIZERS

A unique pelletizing method that handles the low-viscosity materials in an innovative way, known as the Vibrating Ring Droppo Pelletizer, is available. By pumping the material through a small vibrating head, an array of pellet sizes can be produced. The pellet size depends upon the frequency of the head. Typical viscosity range is 100–300 cP. The Droppo method is capable of producing uniform spherical droplets, and offers an alternative for pelletizing low-viscosity melts and liquids that do not form strands. Pellet formation is achieved by the harmonic vibration applied to the melt in a die head. Surface tension causes the flow to break into small droplets. The droplets produced become spherical pellets with an extremely narrow pellet size distribution. There is no mechanical cutting involved. Depending on the viscosity and the surface tension of the melt, drops with diameters from 0.3 to 4.0 mm$\pm$10% are generated by varying the frequency and the size of the die holes (see Figure 7).

After exiting the die, the droplets fall through a cooling tower that uses air, nitrogen, water, or water spray to solidify the product. The height of the tower is dictated by the materials, the throughput rate, and the method of cooling. Such spheres offer the ideal pellet shape for exact dosing and homogenous mixtures, where equal material flow and high-bulk density are required.

Figure 7   Vibrating ring Droppo pelletizing schematic.

V. MICROPELLETIZING

The "standard"-size pellet produced is approximately 3 mm. Larger pellets, up to 20 mm, are possible but uncommon. More typically, the goal is to extrude pellets of 1 mm or smaller, which is preferred for feeding in small single-screw extruders and micromolding machines. All of the pelletizing methods described above may be used to make micropellets, with varying degrees of modifications and probability of success.

For strand micropelletizing, the speed of the pull rolls requires a drive that is independent and can be synchronized with the cutter rpm. The speed of the pull rolls determines the linear speed of the strands; the faster the cutter rpm, the shorter the pellet. Due to geometrical limitations associated with the pull roll/bed knife/cutter assembly, this method has only been successful in producing pellets approximately 0.8 mm in length and larger. Another alternative is to use a flywheel cutter, as described in the "Shape Extrusion" chapter of this book.

For any of the die face pelletizers, the number and the diameter of the holes in the die are dramatically increased. As the pellets are very small (in the 400 μm range), die blockage, material handling, and classification can become challenging. Often times, dies for micropelletization operate at higher pressures than for standard pellets, which may necessitate the use of the gear pump front end attachment to build pressure to the die, particularly for a pelletizing system that is mated to a starve-fed twin-screw extruder.

For low-viscosity carriers, the vibrating ring "Droppo" pelletizer, as described above, might be the ideal choice for micropellets.

VI. DIE DESIGN ISSUES

Die design is the subject of another chapter in this book. Therefore, only an overview specifically relating to the pelletization process is provided. Regardless of the style of pelletizer, the same basic design principles apply.

Pelletizers, just like any other extrusion process, require a quality extrudate. That concept being accepted, the die design becomes a critical part of the total system design. In the case of all dies, there are a few criteria that need to be recognized in order for the die to function properly.

All products do not exhibit the same level of "die swell." The term "die swell" is used when a product increases in diameter as it exits the hole from the high-pressure internal area to the atmospherical area outside the die (see Figure 8).

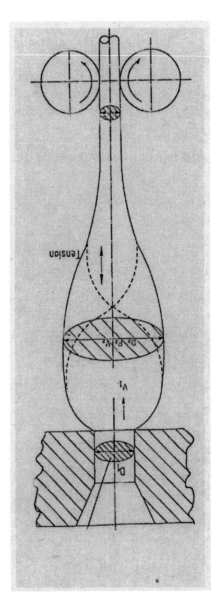

Figure 8 Example of die swell and drawdown.

For a water ring or air quench pelletizer, the die swell calculation is much more critical because a drawdown of the product is not possible. Therefore die swell must be recognized as a critical parameter during the design phase of the die hole geometry to make the correct pellet geometry.

The pressure distribution across the face of the die is a critical design consideration. In a strand die, the comparison of pressure readings from the holes in the center of the die, when compared with the pressure readings at the outside holes, should be held at $\pm 2\%$. A die with a higher differential of pressure will exhibit a higher flow in the center of the die when compared with the outermost strand flow. Given that the feedrolls of the pelletizer will pull all of the strands at the same lineal rate, the outside strands will be a smaller finished diameter than those that were produced by the center of the die. This variation will show up in the final pelletized products as a range of bulk density due to the changes in volume from one pellet to another.

VII. SUMMARY

Pelletization is an established technology in plastics for a wide variety of materials and applications. Accordingly, almost every technical challenge presented by the pharmaceutical industry has been addressed at some level for plastics, and can be successfully applied to manufacture an FDA product. The main issue faced today is for users to convince machine suppliers to downsize existing equipment and to implement the necessary machinery modifications as dictated for use in a Good Manufacturing Practices (GMP) environment. The evolution of continuous processing via single-screw and twin-screw extruders in the medical device and pharmaceutical industries has already made this—to a significant degree—a reality.

REFERENCES

1. Progelhof RC, Throne JL. Polymer Engineering Principles. Cincinnati, OH: Hanser Gardner, 1993:427.
2. Sansone LF. Strand Cooling. Extrusion Solutions v 2.0 CD. SPE Extrusion Division 2001.
3. Dietz W. Polym Eng Sci 1978; 18:1030.
4. Sisfleet WL, Dinos N, Collier JR. Polym Eng Sci 1973; 13: 0.
5. Herrmann H, ed. Granulieren von Thermoplastichen Kunstoffen. VDI-Verlag GMBH.
6. Shoemaker P. Leistritz Training Workshop, June 2002.

10

Melt-Extruded Controlled-Release Dosage Forms

James W. McGinity
The University of Texas at Austin, Austin, Texas, USA

Feng Zhang
PharmaForm L.L.C., Austin, Texas, USA

I. INTRODUCTION

During the past 30 years, extensive research has been pursued with modified- and controlled-release drug delivery systems. The advantages of such systems over traditional dosage forms include improved patient compliance and more constant blood levels of drug, which can significantly increase the efficacy and reduce the side effects of the drug substance [1]. Recent advances in the field of drug delivery have resulted in the precise control of the level and location of drug in the body. In addition, lower doses of medication to the patient are needed [2].

The use of slowly eroding matrix tablets has been one of the most common approaches for preparing controlled-release dosage forms. Water-swellable polymers or polysaccharides have been used as both binders and retardants. The most widely used processes to manufacture rapid and controlled-release tablets include the wet granulation and direct compression techniques. The wet granulation process is both a labor- and equipment-intensive technique involving several steps. Excipients such as binders, lubricants, and glidants are required to facilitate processing. The direct compression and dry granulation techniques are subject to content uniformity and segregation problems during the tableting process. Poor compressibility of the tablet excipients or the drug substance may introduce additional problems into the compaction process.

Hot-melt extrusion technology is one of the most common processing techniques in the plastic industry. For over two decades, the importance of "continuous processing" in the pharmaceutical industry has been recognized. The potential of automation and the reduction of capital investment and labor costs have made hot-melt extrusion worthy of consideration as a pharmaceutical process. The fusion method, which produces granules from a congealed mixture, has been reported to provide slower release profiles compared to the direct compression and wet granulation methods [3]. During melt extrusion, a powder blend of active drug substance, polymer, and excipients is transferred by a rotating screw through the heated barrel of an extruder. The intense mixing and agitation during processing cause suspended drug particles to deaggregate in the molten polymer, resulting in a more uniform dispersion of fine particles. The polymer melts at the elevated temperature and the molten mass is continuously pumped through the die, which is attached to the end of the barrel. The molten polymer or wax component in the formulation rapidly solidifies when the extrudate exits the machine through the die. Depending on the shape of die, the final product may take the form of a film, cylinder, or granule [4].

Until recently, hot-melt extrusion had received limited attention in the pharmaceutical literature. Pellets comprising cellulose acetate phthalate were prepared using a rudimentary ram extruder, and the dissolution rates of active compounds were studied as a function of pellet geometry [5]. More recently, the production of matrices based on polyethylene and polycaprolactone were investigated using extruders of laboratory scale [6,7]. Mank et al. reported on the extrusion of a number of thermoplastic polymers to produce sustained-release granules [8,9]. A melt-extrusion process for manufacturing matrix drug delivery systems was reported by Sprockel et al. [10].

Hot-melt extrusion is a relatively new technique to the pharmaceutical industry, and offers many benefits over traditional processing techniques. The process is anhydrous, which avoids any potential drug degradation because of hydrolysis when aqueous or hydroalcoholic media are employed in the granulation process. In addition, poorly compactible materials can be incorporated into tablets produced by cutting an extruded rod, eliminating any potential tableting problems seen in traditional compressed-dosage forms. As an initial assessment, the thermal, chemical, and physical properties of the drug substance must be characterized. Depending on the physical properties of the drug substance and the other excipients in the formulation, the active moiety may be present as dispersed particles, a solid solution, or a combination in the final dosage form. The state of the drug in the dosage form may have a profound impact on the processability and stability of the finished product.

II. PROCESS AND EQUIPMENT

Hot-melt extrusion equipment consists of an extruder, downstream auxiliary equipment, and other monitoring tools used for performance and product-quality evaluation [11]. The extruder is typically composed of a feeding hopper, barrel, screw, die, screw-driving unit, and a heating/cooling device. Downstream equipment is used to collect the extrudates for further processing. Monitoring devices on the equipment include temperature gauges, a screw-speed controller, an extrusion-torque monitor, and pressure gauges [12].

During the hot-melt extrusion process, different zones of the barrel are preset to specific temperatures before the process is started. A powder blend containing the thermoplastic polymer or retardant waxy material, the drug substance, and other processing aids is fed into the barrel of the extruder through the hopper and transferred inside the heated barrel by a rotating screw. The polymer or waxy retardant material begins to melt or soften once the material enters the compression section of the extruder. Temperatures at different sections of the barrel are normally controlled by electrical heating bands, and the temperature is monitored by thermocouples. The temperature of the melting section is normally set at 30–60 °C above the glass transition temperature of the amorphous polymers or the melting point of semicrystalline polymers. The materials inside the barrel are heated mainly by the heat generated because of the shearing effect of the rotating screw as well as the heat that is conducted from the heated barrel. The molten mass is eventually pumped through the die that is attached to the end of the barrel. The extrudates are subject to further processing by auxiliary downstream devices.

During a continuous extrusion process, the feedstock is required to have good flow properties inside the hopper. For the material to demonstrate good flow, the angle between the sidewall of the feeding hopper and a horizontal line needs to be larger than the angle of repose of the feedstock. In the case of cohesive materials or very fine powders, the feedstock tends to form a solid bridge at the throat of the hopper, resulting in erratic powder flow. Under these conditions, a force-feeding device is sometimes used. Powder properties including bulk density, particle size and shape, and material compactability will influence the flow characteristics of the material in the hopper of the extruder.

The design of the extrusion screw has a significant influence on the efficiency of the hot-melt extrusion process. The function of the screw is to transfer the material inside the barrel and to mix, compress, and melt the polymeric materials and pump the molten mass through the die. Several parameters are used to define the geometrical features of the screw.

Most screws are made from stainless steel, which is surface coated to withstand friction and potential surface erosion or decay that may occur during the extrusion process. Based on the geometrical design and the function of the screw at each section, an extruder is generally divided into three zones: feeding section, melting or compression section, and metering section, as seen in Figure 1. Thermoplastic polymers and waxes exist in a molten state in the metering section. Only single-screw extruders were used during the early days of this technology. Twin-screw extruders are currently used to process larger quantities of material. The two screws can rotate in the same direction (corotating extruder) or in the opposite direction (counterrotating screw). Twin-screw extruders possess many other advantages over single-screw extruders, such as easier material feed, more intensive mixing, less tendency to overheat the materials, and a shorter residence time.

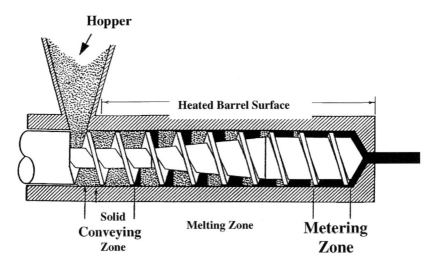

Figure 1 Schematic of an extruder illustrating various functional zones including the hopper, solid conveying zone, melting zone, metering zone, and die.

III. APPLICATIONS TO CONTROLLED-RELEASE ORAL-DOSAGE FORMS

Conventional extrusion/spheronization is an important production process for the preparation of controlled-release pellets, as are solution/suspension techniques [13,14]. The importance of having high-quality pellets or granules for future processing into a pharmaceutical dosage form was recognized by Gamlen [15] as well as Lindberg et al. in the 1980s [16,17]. However, control of porosity, content uniformity, consistent pellet-size distribution, and the design of a truly continuous process were not easily achieved. Pellet technology has advanced in the 1990s with the introduction of new processing equipment. Such developments have given the pharmaceutical scientist numerous opportunities to apply scientific principles to the design of novel dosage forms.

To produce granules or tablets via hot-melt extrusion, a pharmaceutical-grade thermal polymer or lipid material must be selected such that these materials can be processed at a relatively low temperature because of the thermal sensitivity of most drugs. All components must be thermally stable at the processing temperature during the short duration of the heating process. The materials used in the production of controlled-release hot-melt extruded granules and tablets must meet the same level of purity and safety as the excipients that are used in traditional dosage forms. A review of the scientific literature will reveal that the compounds used in the production of hot-melt extruded pharmaceutical products have previously been used in the production of other solid dosage forms such as tablets, pellets, and transdermal films. The materials must be thermally stable and demonstrate acceptable physical and chemical stability.

The functional excipients in the powder blend to be extruded may be broadly classified as matrix carriers, release-modifying agents, bulking agents, and lubricants. The excipients can impart specific properties to hot-melt extruded pharmaceuticals in a manner similar to those in traditional dosage forms. The thermal stability of each individual compound and of the composite mixture should be sufficient to withstand the production process. When the extrudate is cooled to room temperature, the polymer or lipid thermal binder solidifies and bonds the powders together to form a matrix.

The drug-release rate from hot-melt extruded dosage forms is highly dependent upon the characteristics of the carrier material. Several materials that have been incorporated into hot-melt extruded dosage forms are poorly soluble in water [18–21] or have slow hydrating or gelling rates [22,23]. To optimize or regulate the drug-release rate from these systems, functional excipients may be added. Depending upon their physical and chemical properties, various release profiles may be achieved. The inclusion of thermal polymeric carriers usually requires the incorporation of a plasticizer into the formulation in order to improve the processing conditions during the manufacturing of the extruded dosage form and also to improve the stability and physicochemical properties of the final product. The choice of a suitable plasticizer to lower the processing temperature will depend on many factors such as plasticizer–polymer compatibility and plasticizer stability. Triacetin [19], citrate esters [22,24], and low molecular weight polyethylene glycols [19,23,24] have been investigated as plasticizers in hot-melt extruded systems. Plasticizers may also be incorporated into hot-melt extruded dosage forms to improve the physical–mechanical properties of the final dosage form. In transdermal films, the addition of a plasticizer to the polymer matrix can improve the film's flexibility [22,24,25].

Thermal degradation of both the drug substance and the retardant thermal excipient was recognized by Follonier et al. as a limitation of hot-melt processing [18]. Diltiazem hydrochloride, a relatively stable, freely soluble drug was incorporated into their polymer-based pellets for sustained-release capsules. Prior to formulation, polymers and plasticizers were selected to optimize the drug-release profiles. Ethylcellulose (EC), cellulose acetate butyrate (CAB), polyethyl acrylate/methyl methacrylate/trimethyl ammonio ethyl methacrylate chloride (Eudragit RSPM), and polyethylene-*co*-vinyl acetate (EVAC) were the polymers selected for this study, and the plasticizers included triacetin and diethyl phthalate. The porosity of the formulations was assessed by mercury porosimetry. The hot-melt extruded pellets exhibited a smooth surface and low porosity. The in vitro release of diltiazem was biphasic, with the CAB and EVAC pellets giving the slowest release rate, as seen in Figures 2 and 3, respectively. The type and percentage of plasticizer used, drying time of the polymers, extrusion temperatures, and plasticization times varied with each formulation. Adequate stability of the Eudragit RSPM was demonstrated for extrusion at a temperature of 130°C.

Follonier et al. [19] also investigated the influence of additives in several polymeric systems and continued their studies utilizing diltiazem hydrochloride as a model compound. Drug-release profiles from extrudates varying in their polymer-to-drug ratio were analyzed using a double exponential decay equation in order to discriminate between surface release and diffusion-controlled release phases. Pore-forming additives and hydrophilic polymers were incorporated into the formulation in an effort to increase the drug-release rate by increasing the porosity of the pellet during dissolution. Viscosity-inducing agents were incorporated in the polymer matrix to limit the burst

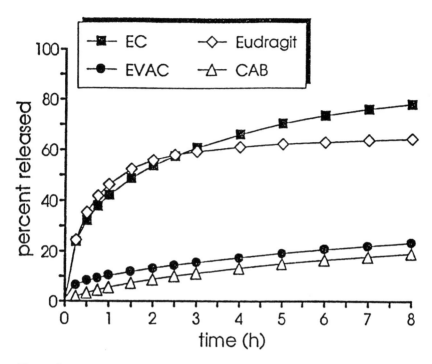

Figure 2 Release profiles of diltiazem hydrochloride from extruded pellets based on various polymers (polymer/drug ratio 1:1, size 2×2 mm). (From Ref. (18).)

effect that is often seen with matrix systems. The incorporation of swelling agents such as AcDiSol and Explotab was also investigated as a method to influence drug release. The release profiles of diltiazem hydrochloride from EVAC-based pellets containing 20% of various swelling agents in pH 7.0 media are seen in Figure 4. The investigators found that the inclusion of enteric polymers, including cellulose acetate phthalate, hydroxypropylmethyl-cellulose phthalate, and Eudragit S, also increased the dissolution rate of diltiazem hydrochloride from EVAC pellets in dissolution media that was maintained at pH 7.0.

The controlled-release properties and mechanism of release of diclofenac sodium from hot-melt extruded wax granules have been reported by Miyagawa et al. [20] and Sato et al. [21]. These researchers utilized a twin-screw compounding extruder to prepare wax matrix granules composed of the model drug, carnauba wax, and other rate-controlling agents. In their first report, they showed that a wax matrix with high mechanical strength could be obtained

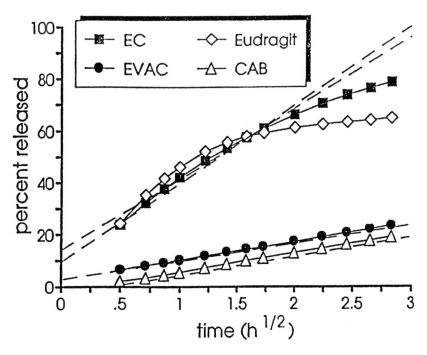

Figure 3 Diltiazem hydrochloride released (%) as a function of the square root of time and calculated regression lines (- - -). (From Ref. (18).)

even at temperatures below the melting point of the wax [20]. The use of wax is advantageous because wax is inert to most pharmaceutical active compounds. Dissolution release profiles of diclofenac sodium from the wax matrix granules were strongly influenced by the formulation of the granules. Hydroxypropyl-cellulose, Eudragit L, and sodium chloride were incorporated into diclofenac sodium/carnauba wax matrices. Increasing the content of either the cellulose derivative or the methacrylic acid copolymer in the wax matrix granules resulted in a substantial increase in the release rate of diclofenac sodium. The release of diclofenac sodium from hydroxypropylcellulose/wax matrices was less pH dependent than that of the system containing wax/Eudragit L because the methacrylic acid copolymer is insoluble in water and in solutions with a pH less than 6.0. The effect of sodium chloride was less pronounced and these results were attributed to the negligible swelling effect of this material. The authors emphasized the advantages of using the twin-screw extruder for wax matrix tablets, such as low temperatures, high kneading and dispersing

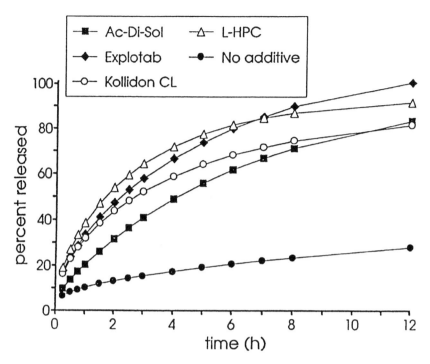

Figure 4   Release profiles of diltiazem hydrochloride from EVAC-based pellets containing 20% of various swelling agents, at pH 7.0. (From Ref. (19).)

ability, and low residence time of materials in the extruder. Sato et al. concluded in the second study that the selection of rate-controlling agents based on physicochemical properties, such as solubility and swelling characteristics, had a significant impact on the properties of wax matrix granules prepared by this extrusion process [21]. The expansion of pores resulting from the presence of the hydroxypropylcellulose in the wax granules causes a structural defect observed as cracking on the surface of the granule. No surface cracks were seen with wax granules containing sodium chloride.

Liu et al. investigated the influence of formulation factors on the physical properties of hot-melt extruded granules and compressed tablets containing wax as a thermal binder/retarding agent [26]. The properties of granules and tablets prepared by melt extrusion were compared with those prepared by a high-shear melt granulation method. Powder blends containing phenylpropanolamine HCl, Precirol® and various excipients were extruded in

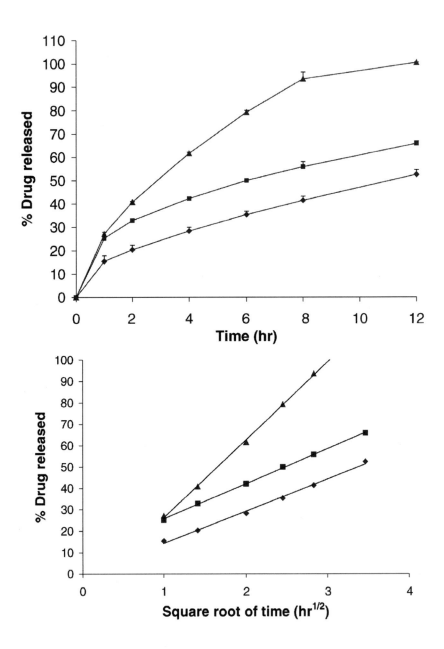

a single-screw extruder at open-end discharge conditions. The extrudates were then passed through a 14-mesh screen to form granules. The extrusion conditions and optimal level of wax needed to function as the thermal binder were dependent on the properties of the filler excipients. At the same wax level, drug release from tablets decreased in order of using microcrystalline cellulose (MCC), lactose, and Emcompress® as the filler excipient, as seen in Figure 5. The dissolution release profiles of phenylpropanolamine HCl from the tablets was due to the differences in the solubility, swellability, and density of the filler excipients. The compression force was shown to have no significant effect on the release rate of the phenylpropanolamine HCl from tablets containing the hot-melt extruded granules, as seen in Figure 6. Replacing Precirol with Sterotex® K, a high-melting-point hydrogenated vegetable oil, resulted in slightly increased dissolution rates when the melt-extrusion process was performed at the same temperature conditions. Hot-melt extruded granules were observed to be less spherical than high-shear melt granules and showed lower values of bulk/tap densities. However, tablets containing MCC or lactose granules prepared by hot-melt extrusion exhibited higher hardness values. Slower drug-release rates from tablets were found for granules containing MCC that were prepared by hot-melt extrusion compared to melt granulation. Analysis of hot-melt extruded granules showed better drug content uniformity among granules of different size ranges compared to high-shear melt granules, resulting in a more reproducible drug-release profile from the corresponding tablets.

Other materials may also be included in the formulation of hot-melt extruded dosage forms such as thickening agents and antioxidants. Cuff and Raouf [27] investigated the influence of adding microcrystalline cellulose into PEG 8000 matrices in order to improve the formulation viscosity and the plasticity of the resulting tablets formed by injection molding. They reported that fenoprofen calcium inhibited the hardening of a PEG–MCC matrix, resulting in an unusable product.

Excessive temperatures needed to process unplasticized or underplasticized cellulose-based polymers (hydroxypropylcellulose or ethylcellulose)

Figure 5 Influence of filler excipient on the release of PPA from tablets containing hot-melt extruded granules using MSP Method I at 100 rpm in 500 ml purified water, maintained at 37°C. ▲: 15% PPA, 30% Precirol, and 55% MCC; ■: 15% PPA, 30% Precirol, and 55% lactose; ◆: 15% PPA, 30% Precirol, and 55% Emcompress. Top: % Drug released vs. time. Bottom: % Drug released vs. square root of time. (From Ref. (26).)

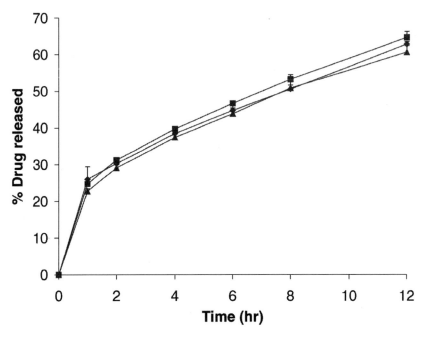

Figure 6 Influence of compression force on the release of PPA from tablets containing hot-melt extruded granules using USP Method I at 100 rpm in 500 ml purified water, maintained at 37°C. The tablets were composed of 15% PPA, 30% Precirol, and 55% lactose compressed at ■: 1000 kg, ◆: 2000 kg, ▲: 3000 kg. (From Ref. (26).)

may lead to polymer oxidation. One manufacturer of these materials recommends the incorporation of an antioxidant such as butylated hydroxytoluene or ascorbic acid into formulations containing low molecular weight hydroxypropylcellulose [28]. Similarly, a combination of an antioxidant, light absorber, and acid acceptor is recommended for systems employing ethylcellulose [29]. Repka and McGinity reported that vitamin E TPGS (TPGS, D-α-tocopheryl polyethylene glycol 1000 succinate) in a HPC/PEO 50:50 ratio film, decreased the degradation and chain scission of the polymer blend during the melt-extrusion process, and stabilized the polyethylene oxide that is susceptible to thermal oxidation [30].

Thermal stability of the individual compounds in hot-melt extruded granulated tablets is a prerequisite for the process, although because of the short processing times not all thermolabile compounds are excluded. The incorporation of plasticizers may lower the processing temperatures required in hot-melt extrusion, thereby reducing drug and carrier degradation. Drug release from these systems can be modified by the incorporation of various functional excipients. The dissolution rate of the active compound can be increased or decreased, depending on the properties of the dissolution-rate-modifying agent. For systems that display oxidative or free-radical degradation during processing or storage, the addition of antioxidants, acid acceptors, and/or light absorbers may be warranted.

Another application of hot-melt extrusion was described by Zhang and McGinity [23]. These researchers investigated the properties of polyethylene oxide (PEO) as a drug carrier and studied the release mechanism of chlorpheniramine maleate (CPM) from matrix tablets. In these extruded tablets, PEG 3250 was included as a plasticizer to facilitate processing. The stability of the PEO as a function of processing temperature was reported. The polymer type, temperature, and residence time in the extruder was shown to be of great importance. These authors reported that the drug, polymer, and other ingredients must be stable at the elevated processing temperature during the approximate 2 min that the powder blend is processed through the equipment. Additional mixing of the components occurred in the barrel of the extruder, because the content uniformity of the extruded tablets was within 99.0% to 101.0% of the theoretical content. During the hot-melt extrusion process, a dry powder blend of drug, polymer, and other adjuvants was fed into the extruder and softened inside the barrel of the machine. The molten mass was extruded through a rod-shaped die and then cut manually into 400-mg tablets. CPM and PEO were shown to be stable under the processing conditions. The molecular weight of the PEO, the drug-loading percentage, and the inclusion of polyethylene glycol as a processing aid were all found to influence the processing conditions and the drug-release properties of the extruded tablets. Drug release from the matrix tablet was controlled by erosion of the PEO matrix and the diffusion of the drug through the swollen gel layer at the surface of the tablets. CPM was dispersed at the molecular level in the PEO matrix at low drug-loading level and recrystallization of CPM was observed at high drug-loading levels.

The profiles in Figure 7 illustrate the influence of PEG 3350 on the release of CPM from hot-melt extruded matrix tablets. It can be seen that as the percentage of PEG 3350 increases, the release rate of CPM increased. The PEG 3350 hydrated and dissolved faster than the PEO. The hydration and dissolution rate of the entire matrix system were thus accelerated because of the presence of the PEG. The presence of the PEG also decreased the viscosity of the hydrated gel layer, which facilitated the diffusion of the CPM from the swollen gel layer surrounding the tablets.

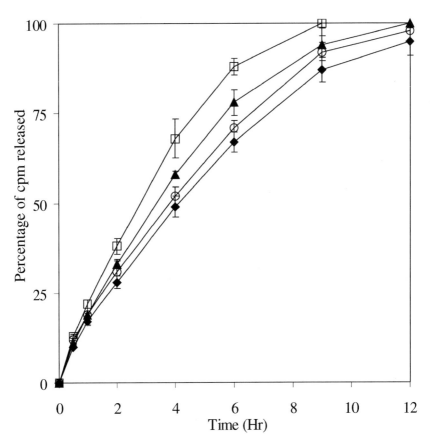

Figure 7 Influence of PEG (3350) on the release of chlorpheniramine maleate from matrix tablets using USP Method II at 37°C and 100 rpm in 900 ml purified water. ♦: 6% CPM, 0% PEG (3350), and 94% PEO (1.0 m); O: 6% CPM, 6% PEG (3350), and 88% PEO (1.0 m), ▲ 6% CPM, 20% PEG (3350), and 74% PEO (1.0 m); □, 6% CPM, 40% PEG (3350), and 54% PEO (1.0 m). (From Ref. (23).)

The release of CPM from the matrix tablet as a function of the dissolution media is shown in Figure 8. Faster drug release from the tablets was seen in acidic medium (0.1 N HCl) compared to purified water or phosphate buffer (pH 7.4, 50 mM). The solubility of CPM at 37°C in all three media exceeds 1 g/ml. The difference in the dissolution profiles could be the result of an interaction between PEO and the hydrogen ions in the acidic medium. The ether oxygen atom of PEO has two spare pairs of electrons, and

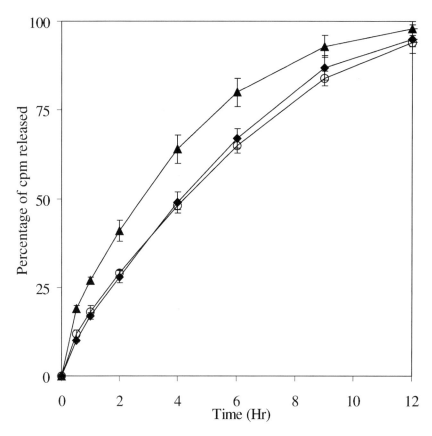

Figure 8 Influence of dissolution medium on the release of CPM from matrix tablets (6% CPM, 6% PEG, and PEO 1.0 m) using USP Method II at 37°C and 100 rpm in 900 ml dissolution medium. ▲: 0.1 N HCl without enzyme; ◆: purified water; ○: simulated intestinal fluid (pH 7.4) without enzyme. (From Ref. (23).)

it is able to form a hydrogen bond with the abundantly presented hydrogen ions in the acidic media. The interaction with the hydrogen ions would result in the strong electrostatic repulsion between the polymer chains. This thermodynamically favored process increases the solubility and dissolution rate of PEO in the acidic media, resulting in the faster drug-release rate. Bailey et al. [31] also found that the PEO was more rapidly soluble in acidic medium and observed that the upper critical temperature of an aqueous PEO solution was significantly raised in acidic media by a strong hydrogen bonding effect.

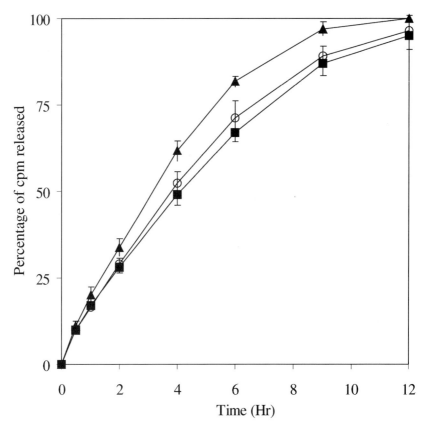

Figure 9 Influence of drug loading on the release of CPM from matrix tablets using USP Method II at 37°C and 100 rpm in 900 ml purified water. ■: 6% CPM and 94% PEO (1.0 m); ○: 12% CPM and 88% PEO (1.0 m); ▲: 20% CPM and 80% PEO (1.0 m). (From Ref. (23).)

The influence of drug loading on the release of CPM is shown in Figure 9. When the drug content was increased from 6% to 12%, no change in the percentage of drug release with respect to time was observed. There was only a slight increase when the drug loading reached 20%. The study conveys the reproducibility of dissolution data for tablets produced by hot-melt extrusion. The profiles in Figure 10 demonstrate that drug release was significantly reduced when the PEO (7 m) was present in the polymeric matrix.

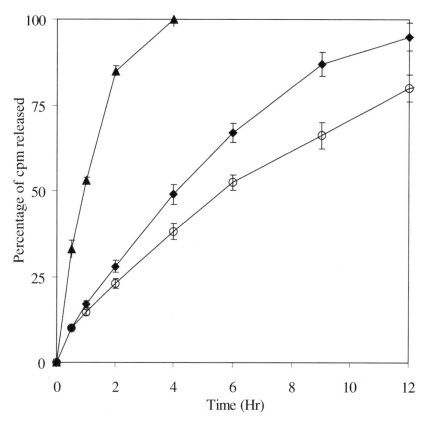

Figure 10 Influence of molecular weight of PEO on the release of CPM form matrix tablets using USP Method II at 37°C and 100 rpm in 900 ml purified water. ▲: 6% CPM, 20% PEG (3350), and PEO (100,000); ◆: 6% CPM, 20% PEG (3350), and PEO (1.0 m); ○: 6% CPM, 20% PEG (3350), and PEO (7.0 m). (From Ref. (23).)

Since both the drug and the polymer must remain stable during processing, polymeric materials suitable for thermal processing require either a low glass transition temperature, or a low melting point in the case of semicrystalline polymers. Poly(vinyl acetate) is a homopolymer synthesized from vinyl acetate monomers via a free radical polymerization technique. It is amorphous because of the presence of an acetate ester side chain in the backbone structure. The glass transition temperature of poly(vinyl acetate) is relatively low because of its highly flexible backbone structure, as seen in Figure 11. The glass transition temperatures of two grades of PVAc of

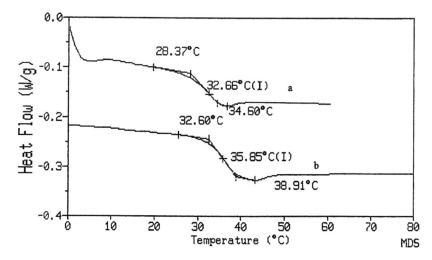

Figure 11 Differential scanning calorimetry of poly(vinyl acetate). (a) Sentry Plus 12 (MW 12,000). (b) Sentry Plus 40 (MW 40,000). (From Ref. (32).)

molecular weight 12,000 and 45,000 were 32.7° and 35.9°C, respectively [32]. Although the polymer is insoluble in water, it is slightly hydrophilic and is able to absorb water to a slight extent. The polymer has been used in the preparation of matrix pellets [33], sustained-release coatings [34], and buccal drug delivery systems [35].

The influence of theophylline loading on the release properties of tablets compressed from extruded granules containing poly(vinyl acetate) is shown in Figure 12 [32]. Because of the low glass transition temperature of the PVAc, the hot-melt extrusion processing temperature was approximately 70°C. The PVAc was demonstrated to have a high solids-carrying capacity when processed by hot-melt extrusion. A powder blend containing 50% theophylline could be readily processed. When drug loading was below 25%, less than 20% of the theophylline was released from the matrix after 12 hr and two distinctive phases were seen in the dissolution profiles (Figure 12). From percolation theory, only finite clusters that were isolated from the surrounding dissolution medium were formed when drug loading was below the percolation threshold. The initial fast release rate was because of the small percentage of finite clusters that was located on the surface of the matrix and connected to the dissolution medium. As demonstrated by the drug-release profiles in Figure 12, the percolation threshold for theophylline was approximately 20%. For granules containing higher levels of theophylline, clusters were

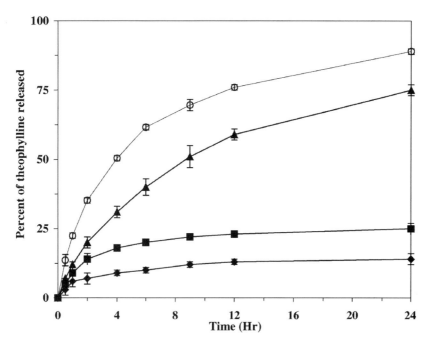

Figure 12 Influence of drug loading levels on release of theophylline from tablets containing hot-melt extruded granules, in 900 ml purified water at 37°C and 100 rpm using the USP Method II (*n*=3). Tablet formulation: 20%, 300 to 425 μm extruded theophylline granules [(♦) 5%; (■) 15%; (▲) 25%; and (○) 50%); 2% PEG and PVAc qs to 100%], 79.5% Avicel® PH 200, and 0.5% magnesium stearate. (From Ref. (32).)

more extensive and the matrix was less tortuous. This contributed to faster drug-release rates at higher drug-loading levels.

As shown in Figure 13, there was no significant difference in the release rates of theophylline in purified water at 100 rpm from tablets containing hot-melt extruded granules at the 10% and 20% loading levels. A further increase in the level of the granules resulted in a decrease in drug-release rate. At low granule loading in the tablets, hot-melt extruded granules were separated from each other inside the matrix tablets. As the concentration of PVAc granule in the tablets increased, the polymeric particles fused together during the compression process because of the low glass transition temperature and hydrophobicity of the PVAc. The fused granules formed hydrophobic domains that were larger than the individual granules. Drug release from these domains

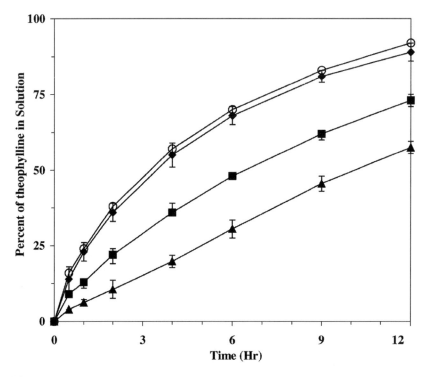

Figure 13 Influence of the loading level of hot-melt extruded granules on the release of theophylline from tablets in 900 ml purified water at 37°C and 100 rpm using the USP Method II ($n = 3$). Tablets: 180- to 212-μm extruded granules (25% theophylline, 2% PEG, and 73% PVAc), 0.5% magnesium stearate and Avicel® PH 200 q.s. to 100%; (O) 10% granule loading; (♦) 20% granule loading; (■) 30% granule loading; (▲) 60% granule loading. (From Ref. (32).)

was significantly slower than from the individual particles because of the resulting increase in tortuosity.

Scanning electron microscopy was used to investigate the surface structure of the PVAc matrix tablets prior to and following the dissolution test. The surface morphology of the hot-melt extruded tablets with and without water-soluble polymers is shown in Figures 14 and 15, respectively. A smooth surface was observed for the tablets containing PVAc after the dissolution test. PVAc was in a rubbery state at 37°C, and the smooth surface was the result of the hydrodynamic shearing force imposed on the surface of the tablet by the

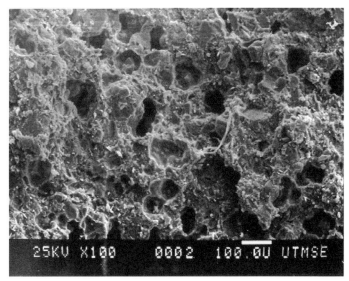

Figure A

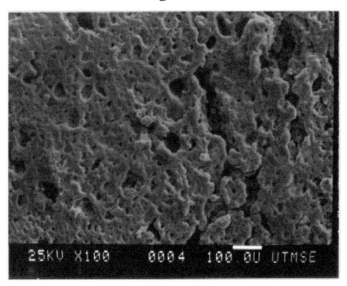

Figure B

Figure 14 Surface morphology of hot-melt extruded tablets. Tablet (25% theophylline, 25% lactose, 20% PEO 1,000,000, and 30% PVAc): (A) prior to dissolution test; (B) following 12-hr dissolution test. (From Ref. (32).)

Figure A

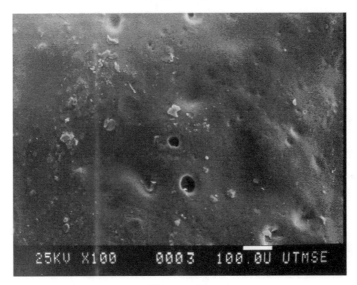

Figure B

Figure 15 Surface morphology of hot-melt extruded tablets. Tablet (25%
theophylline, 45% lactose, and 30% PVAc): (A) prior to dissolution test; (B)
following 12-hr dissolution test. (From Ref. (32).)

agitated dissolution medium. Pores on the tablet surface resulted from the capillary diffusion of the theophylline and lactose. The surface of the extruded tablets containing PEO 300,000 prior to the dissolution test was significantly rougher. PEO is a water-soluble polymer and is not miscible with PVAc. Hot-melt extruded matrix tablets were therefore less cohesive when PEO was added. As seen in Figure 14, the surface of the PVAc matrix tablets with PEO was porous following the dissolution test when examined by a scanning electron microscope. Following the completion of the dissolution study, the hot-melt extruded tablets that contained PEO were freeze-dried, and the weight of the tablets was found to be equal to the theoretical weight of the PVAc in the tablets. The water-soluble polymer was demonstrated to be more effective in facilitating the drug-release process because the tablets were more porous, and the drug was able to diffuse through the hydrated gel layer during the dissolution process. With lactose, the dissolution and movement of drug molecules through the capillary channels could be hindered by the dissolution of the lactose. Drug release was slower when PEO (MW 1,000,000) was used in the extrudate because both the hydration and dissolution of the PEO were slower, and the diffusivity of theophylline in the polymeric hydrogel also influenced the dissolution process.

Poly(vinyl acetate) was demonstrated to be an excellent carrier for the preparation of controlled-release granules processed by a hot-melt extrusion. Because of the low glass transition temperature of the polymer, the melt extrusion process could be conducted at temperatures within the range of 50° to 70°C. During processing, the extrudate was subjected to minimal thermal and mechanical stress. In the absence of water-soluble polymers as drug-release modifiers, the extrudates had to be ground into fine powder and compressed into tablets with directly compressible excipients to achieve desirable drug-release profiles. Water-soluble polymers, such as PEO and HPMC, were shown to facilitate the drug release to a much greater extent than lactose. Tablets with desirable release properties could be made by cutting the extrudate when water-soluble polymers were added as drug-release modifiers. Theophylline was released from the melt-extruded systems by diffusion. Drug-release data for the poly(vinyl acetate) matrix tablets were in good agreement with the Higuchi model and the percolation theory [31].

IV. SUMMARY

Hot-melt extrusion technology facilitates the design and development of controlled-release oral-dosage forms without the use of water or solvents as a granulating medium. Materials that melt or soften at the processing temperature function as both binding agents and controlled-release agents to modulate the release of the active moiety from the resulting granules or extruded tablets. Hydrophilic and hydrophobic retardants can be present in the powder blend and only one component in the formulation must soften or melt in order to form the extruded product. Drug substances may also dissolve in the molten mass to form a molecular dispersion.

Melt extrusion technology is suitable for both high-dose and potent compounds. The mixing that occurs in the barrel of the extruder during processing ensures good content uniformity of the active material in the finished product. All components in the powder blend, however, must be stable at the processing temperature and also during the short period that the powder is exposed to the elevated temperatures.

Hot-melt extrusion technology holds great promise for the future as the process of choice for the design and development of controlled-release oral-dosage forms.

REFERENCES

1. Welling PG, Dobrinska MR. Controlled Drug Delivery. New York: Marcel Dekker, Inc., 1987.
2. Langer R. Drug delivery and targeting. Nature April 30, 1998; 392(suppl 5):5–10.
3. Saraiya D, Bolton S. The use of Precirol® to prepare sustained release tablets of theophylline and quinidine gluconate. Drug Dev Ind Pharm 1990; 16(13): 1963–1969.
4. Aitken-Nicol C, Zhang F, McGinity JW. Hot melt extrusion of acrylic films. Pharm Res 1996; 13(5):804–808.
5. Rippie EG, Johnson JR. Regulation of dissolution rate by pellet geometry. J Pharm Sci 1969; 58:428–431.
6. Shivanand P, Hussain AS, Sprockel DL. Factors affecting release of KCl from melt extruded polyethylene disks. Pharm Res 1991; 8:S-192.
7. Prapaitrakul W, Sprockel DL, Shivanand P, Sen M. Development of a drug delivery system through melt extrusion. Abstracts of the 4th Am. Assoc. Pharm. Scientists, Atlanta, 1989. Pharm Res 1989; 6:S-98.
8. Mank R, Kala H, Richter M. Darstellung wirkstoffhaltiger Extrusionformlinge auf der Basis von Thermoplasten, Teil 1, Untersuchungen zur Wirkstoffliberation. Pharmazie 1989; 44:773–776.
9. Mank R, Kala H, Richter M. Darstellung wirkstoffhaltiger Extrusionformlinge auf der Basis von Thermoplasten, Teil 2, Untersuchungen zur Optimierung der Wirkstoffreigabe. Pharmazie 1990; 45:592–593.

10. Sprockel O, Sen M, Shivanand P, Prapaitrakul W. A melt-extrusion process for manufacturing matrix drug delivery systems. Int J Pharm 1997; 155:191–199.

11. Kruder GA. Extrusion, In: Kroschwitz J, ed. Encyclopedia of Polymer Science and Engineering. Vol. 1. 2d ed. New York: John Wiley & Sons, Inc., 1985: 571–631.

12. McGinity JW, Koleng JJ, Repka MA, Zhang F. Hot melt extrusion technology. In: Encyclopedia of Pharmaceutical Technology. Vol. 19. 2nd ed. New York: Marcel Dekker, Inc., 2000:203–226.

13. Reynolds A. A new technique for the production of spherical particles. Manuf Chem 1970; 41:40–44.

14. Harris MR, Ghebre-Sellassie I. In: McGinity JW, ed. Aqueous Polymeric Coatings for Pharmaceutical Dosage Forms, Chap. 3. New York: Marcel Dekker, Inc, 1997:81–100.

15. Gamlen M. Continuous extrusion using a Baker Perkins MP50 (multipurpose) extruder. Drug Dev Ind Pharm 1986; 12:1701–1713.

16. Lindberg NO, Tufvesson C, Olbjer L. Extrusion of an effervescent granulation with a twin screw extruder, Baker Perkins MPF 50 D. Drug Dev Ind Pharm 1987; 13:1891–1913.

17. Lindberg NO, Tufvesson C, Holm P, Olbjer L. Extrusion of an effervescent granulation with a twin screw extruder, Baker Perkins MPF 50 D. Influence on intragranular porosity and liquid saturation. Drug Dev Ind Pharm 1988; 14:1791–1798.

18. Follonier N, Doelker E, Cole ET. Evaluation of hot-melt extrusion as a new technique for the production of polymer-based pellets for sustained-release capsules containing high loading of freely water soluble drugs. Drug Dev Ind Pharm 1994; 20(8):1323–1339.

19. Follonier N, Doelker E, Cole ET. Various ways of modulating the release of diltiazem hydrochloride from hot-melt extruded sustained-release pellets prepared using polymeric materials. J Control Release 1995; 36:243–250.

20. Miyagawa Y, Okabe T, Yamaguchi Y, et al. Controlled release of diclofenac sodium from wax matrix granule. Int J Pharm 1996; 138:215–224.

21. Sato H, Miyagawa Y, Okabe T, et al. Dissolution mechanism of diclofenac sodium from wax matrix granules. J Pharm Sci 1997; 86(8):929–934.

22. Repka MA, Gerding TG, Repka SL, McGinity JW. Influence of plasticizers and drugs on the physical–mechanical properties of hydroxypropylcellulose films prepared by hot-melt extrusion. Drug Dev Ind Pharm 1999; 25(5): 625–633.

23. Zhang F, McGinity JW. Properties of sustained release tablets prepared by hot-melt extrusion. Pharm Dev Technol 1998; 14(2):241–250.

24. Aitken-Nichol C, Zhang F, McGinity JW. Hot-melt extrusion of acrylic films. Pharm Res 1996; 13:804–808.

25. Repka MA, McGinity JW. Physical–mechanical moisture absorption and

bioadhesive properties of hydroxypropylcellulose hot-melt extruded films. Biomaterials 2000; 21(14):1509–1517.

26. Liu J, Zhang F, McGinity JW. Properties of lipophilic matrix tablets containing phenylpropanolamine hydrochloride prepared by hot-melt extrusion. Eur J Pharm Biopharm 2001; 52(2):181–190.

27. Cuff G, Raouf F. A preliminary evaluation of injection molding as a technology to produce tablets. Pharm Technol June 1998; 22(6):97–106.

28. Klucel® Hydroxypropylcellulose; Physical and Chemical Properties. Technical Bulletin 250-2W. Wilmington: Hercules, Inc., 1997.

29. Aqualon® Ethylcellulose; Physical and Chemical Properties. Technical Bulletin 250-42. Wilmington: Hercules, Inc., 1996.

30. Repka MA, McGinity JW. Influence of Vitamin E TPGS on the properties of hydrophilic films produced by hot-melt extrusion. Intl J Pharm 2000; 202: 63–70.

31. Bailey FE, Callard RW. Some properties of poly (ethylene oxide) in aqueous solution. J App Polym Sci 1959; 1(1):56–62.

32. Zhang F, McGinity JW. Properties of hot-melt extruded theophylline tablets containing poly (vinyl acetate). Drug Dev Ind Pharm 2000; 29(6):938–948.

33. Schmidt WG, Mehnert W, Fromming KH. Controlled release from spherical matrices prepared in a laboratory scale rotor granulator-release mechanism interpretation using individual pellet data. Eur J Pharm Biopharm 1996; 42: 348–350.

34. Batra V, Bhowmick A, Behera BK, Ray AR. Sustained release of ferrous sulfate from polymer-coated gum arabica pellets. J Pharm Sci 1994; 83:632–635.

35. Poly (vinyl acetate) product brochure, UC-1166A. Danbury, CT: Union Carbide Corporation, 1996.

11
Shape Extrusion

Bob Bessemer
The Conair Group, Inc., Pittsburgh, Pennsylvania, USA

I. INTRODUCTION

Shape extrusion is when the process melt is directly extruded into a part with specific dimensions, and either cut to a length or wound. The extrudate can be a simple rod, or complex shape, referred to as a "profile" (see Figure 1). As biopolymers have been introduced, process and equipment enhancements have made extrusion more of a science. A better relationship has been established between the variables of the process and the potential effects not only on size and shape, but on many physical properties as well. Much of this science can be used when processing active biomedical ingredients into shapes.

The extruder used for pharmaceuticals is often a twin screw due to its ability to mix and convey solids, liquids, or gases. It is imperative that the extruder provides a mixed, devolatilized, and thermally homogenous melt stream at a constant pressure to the die. At this point, the job of the downstream equipment is to form the part.

Dies for pharmaceuticals, the subject of another chapter in this book, often operate at lower temperatures than conventional plastic processes, and do not necessarily need heaters, such as those used for plastics. In fact, it actually might have cooling passages for water or gas (nitrogen, etc.) to help stiffen the extrudate to maintain the shape upon exiting the die. In some cases, dies for pharmaceuticals might not only use specialized tool steels, but also use coatings or even inserts to minimize sticking. The extrudate is then pulled through a cooling medium, and cut or wound as required.

Figure 1 Example of extruded shapes.

A careful consideration is required to properly match the extruder screw design to the material to offer optimum pumping without causing unwanted material issues, such as burning or melt fracturing. The die size, associated geometry, materials of construction, and even machining practices are "material-specific."

Most shape extrusions, whether for plastics or pharmaceutical products, are pulled through the cooling medium by a pulling device. These devices are responsible for the consistent drawing or pulling of the extrudate from the extrusion die to maintain consistent size.

To further discuss the extrusion of pharmaceuticals, it is necessary to understand the extrusion of plastics and rubber. This basic knowledge will clarify how the process and the equipment are modified to facilitate the extrusion of shapes for pharmaceutical applications.

II. COOLING AND CONVEYING PHARMACEUTICAL EXTRUSIONS

A. Free Extrusion Versus Vacuum Sizing

The basic "free" extrusion process, where the material is cooled in air or water without the use of vacuum, can be used to extrude many shapes and materials,

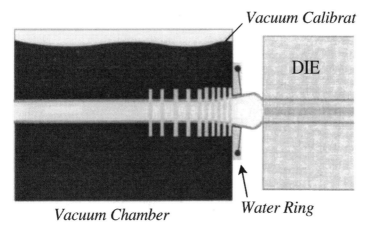

Figure 2   General vacuum calibration.

whether solid or hollow. An extrusion die or a material-shaping fixture is located at the exit of the extruder, similar to the pelletizing process. Depending on the material and throughput rate, the die shape will be somewhat larger than the final product, and will enable "drawdown," as most extrusions are pulled from the die. In other words, the die opening is somewhat larger than the final product dimensions.

Calibration tooling is often utilized for complex shapes so that a vacuum-generating device, such as a liquid ring vacuum pump or regenerative vacuum blower, creates a cooling chamber with lower than atmospherical pressure (see Figure 2). With vacuum applied to a substantially closed-to-atmosphere cooling tank, the pressure differential between the tank and the inside of a hollow extrusion can be used to exert an outward pressure. The profile passes through this calibration tooling after exiting the die, and while in the molten state, the applied vacuum holds the hollow extrusion in its desired shape while it cools. The length of the tool depends on the material and the production rates. The vacuum level is either manually controlled or automatically controlled through a feedback device, such as a laser gauge measuring the outside dimensions of the actual product.

B.   Cooling/Sizing in a "Dry" Environment

If the extrudate does not need to be drawn from the die for sizing, a simple conveyor can be used to support and to convey the part from the die. For years in the plastic industry, air racks were used to cool profiles, especially in small production runs. These cooling tables, which typically are 1–10 m in length, can utilize a series of forming guides to support the profile as it is cooled. A series of fans and air nozzles can be used to direct the air at the extrudate to enhance cooling. These air-cooling tables enable a more gradual cooling rate for a controlled temperature transfer across the profile.

The most basic cooling or drying method for pharmaceutical products after exiting the extruder die is a belt-type conveyor (see Figure 3). This conveyor can utilize a solid belt of a Food and Drug Administration (FDA)-approved material, or be more of a screen-type belt, to allow drying and/or cooling from all sides. The construction materials of the screen would be stainless steel or plastic for ease of washdown. A conveyor can use air-cooling fans both above and below the extrudate for cooling and/or drying the part prior to cutting or further processing. A heating hood can be integrally mounted with the conveyor to enhance the drying or processing of the material, if necessary. The speed of the conveyor must be matched to the flow of the profile. In some cases, a loop control device, whether dancer-type or ultrasonic-type, senses the loop of the extrudate and outputs a signal to control the speed of the conveyor. In this way, even very delicate materials can be conveyed without distortion or breakage. The length of this conveyor depends on the rate of production and the material properties.

An example of conveyor cooling might be a pharmaceutical extrusion operation that requires a twin-screw extruder to mix the active ingredient within a bar or shape. A conveyor matches the line speed to support the part while it is cooling, or, in some cases, dried prior to the next operation.

Dry vacuum calibration might be used for complex shapes if the pharmaceutical extrusion material cannot contact water (see Figure 4). The

Figure 3   Close-up belt conveyor.

Figure 4   Example of dry calibration tooling.

vacuum calibration tool is designed with both internal water cooling and vacuum passages incorporated into the design. In this style of tooling, the cooling coolant passes through the calibration tool and never contacts the extrudate. The vacuum calibrator is typically constructed of stainless steel. The tooling will be chilled or heated via liquid to tailor the cooling and/or drying rates required for the materials. The controls associated with the calibration tool include the vacuum level, the temperature, and the number of calibration zones. Downstream of the calibration tool, a conveyor might still be used to offer support and as an area for additional cooling or drying as dictated by the materials being processed.

C.   Cooling/Sizing in Liquid

In plastics extrusion, it is very typical to use a cooling tank that uses water as a cooling medium (see Figure 5). The extrudate is either immersed within the tank or sprayed with a fine, atomized mist to enhance cooling rates. The length of these cooling tanks is dependent on the rate of extrusion and how quickly the extrudate will release the BTUs (British Thermal Units) inherent to the extrudate.

In some medical extrusion processes, deionized water is used within the cooling tank to minimize contamination. The tank is constructed of stainless steel, including all fittings and valves, as the deionized water will quickly

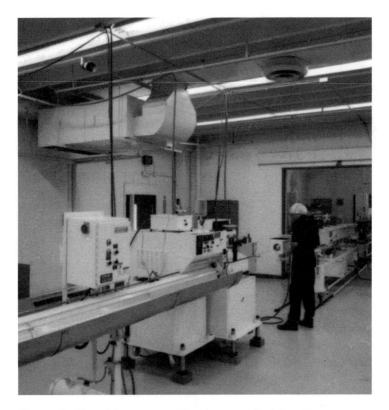

Figure 5 Extrusion system with water trough, belt puller, rotary cutter, and conveyor.

corrode bronze. In some applications involving water-soluble extrusions, a brine solution has been used in place of water, which enables adequate cooling rates without degrading the material.

A closed-loop water circulation system is used to maintain consistent water temperature within the cooling tank. A water pump with a stainless steel housing and an impeller is typically used in combination with filtration capable of removing pyrogens (dead germ bodies) from the water. The filtered water then passes through a stainless steel heat exchanger connected to a chiller system. The chiller system, which does not make contact with the process water, is specified based on the heat being removed.

The "coolant" in the tank can also be heated to facilitate a more gradual cooling process. As with the chiller, these units are totally separate from the process water in that a heat exchanger and a modulating valve are used to automatically regulate the water being brought into the tank. An immersion heater is used within the unit to heat the water. A thermocouple is typically mounted in the cooling or vacuum tank. A programmable logic controller (PLC) controls the chiller or heater unit to maintain the set point temperature.

If the extrudate is hollow and is able to make contact with water, "wet" vacuum sizing tanks may be used. As with dry calibration, a vacuum would be used to apply an outward force to the hollow part, with the calibration tool being the growth-limiting device while the material is setting up.

III. TAKEOFF DEVICES TO PULL THE MATERIAL THROUGH THE COOLING MEDIUM

If the extrudate is being drawn from the extrusion die, the takeoff is a critical component in maintaining product size, regardless of the type of unit. In addition to a precision mechanical design, the motor drive, discussed in another chapter, is extremely important. For pharmaceutical applications, a servodrive is typically utilized to direct drive the belts or wheels with minimal linkage required. A sealed-for-life in-line planetary reducer is used if further torque or speed reduction is required. Servodrives utilize a serial operator interface to control speed and/or position.

Figure 6 Combination pinch roll-cutter.

The most basic takeoff would be a roller or multiple of rollers used above the conveyor to hold the pharmaceutical profile against the conveyor. This option offers a minimal amount of pulling force. Obviously, these rollers would be made of stainless steel, or another FDA-approved material.

Wheel pullers have historically been used for applications in the extrusion of plastics and rubber, where only two points of contact are required with minimal pulling force (see Figure 6). Most applications using wheel pullers involve small-diameter flexible medical tubing. The pinching action of the rolls can be used to maintain an internal air pressure for a tube. Alternatively, a wheel puller, referred to as a "tire puller" with under-

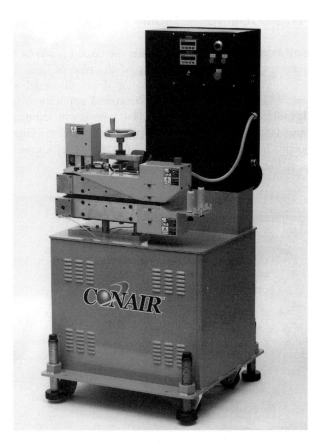

Figure 7   Example of belt puller.

inflated pull rolls, can be specified to slightly wrap around the product for better traction.

Belt pullers are the most common devices used to pull the extrudate from the die and through the cooling medium (see Figure 7). Belt pullers consist of an upper and a lower boom or conveyor track, which, when closed on the profile, applies traction and pulling force. The length of these belts varies depending on the pulling force required and the potential of the extrudate to deform. Typical units utilize belts that are 150–300 mm in length; shorter or longer belts are also available. Poly-V, timing belts, and flat belts are all available and supplied with FDA-approved materials, such as natural rubber and nitriles of various durometers. In some cases, high-temperature or low-sticking belt coverings such as silicone are used. Belts can also be contoured to shape a product during the pulling phase, if amenable.

IV. CUT-TO-LENGTH AND WINDING DEVICES

If the pharmaceutical extrusion needs to be cut to a predetermined length, this is accomplished in-line with several pieces of equipment, including traveling shears, traveling saws, and rotary knife cutters. The application of these devices depends on line speed, cuts per minute, and material.

A. Shears, Saws, and Cutters

Guillotine or shear cutters are relatively slow cutting devices that use a blade, shearing the part off against a bed knife (see Figure 8). It is very typical to use an air cylinder to cycle the blade up and down for the cutting sequence, which results in a momentary product stoppage or interruption. Due to stoppage during the cutting cycle, these cutters may be installed on a traveling table to minimize interruption. These units are also limited to about 15–20 cuts/min, precluding higher line speeds and shorter cut lengths. Shears tend to use relatively thick blades, which limit cutting applications to materials that can be displaced by the blade while not fracturing. The input signal for cut activation can be a simple microswitch, timer, counter, or photo eye.

Traveling saws are commonly used as an in-line cutoff device for rigid plastics. A traveling bed with a clamp is secured to the extrudate while cutting. Upon a signal from an input device, the traveling bed can be made to track the line speed by either pneumatics or electronic-driven linear actuators. Once the table is traveling with the extrudate, the clamp is activated to secure the extrudate while the blade saws through the part. Depending on the style of

Figure 8 Shear cutter.

the saw, the blade can cut up through the table (up cut), down on an angle (chop saw), or across the part (cross cut). Typically, these in-line travelling saws are limited to up to 25 m/min with up to 15–25 cuts/min. In pharmaceutical applications, a toothless blade, similar to a meat-cutting blade, might be used to eliminate the creation of particulates.

Rotary knife cutters are popular for cutting rubber and plastics, especially with the introduction of servomotor drive technology (see Figure 9). The ability to cut simply, cleanly, and quickly makes these units the ideal fit for many pharmaceutical applications. A blade is mounted to a disc, which is driven by a servomotor. Cutter bushings, which are designed to support the extrudate, are mounted such that the blade passes between bushings, allowing the extrudate to be sheared off between them (see Figure 10). The bushings and all components coming in contact with the product are constructed of stainless steel. An advantage of this design is that instead of the material being removed as with a toothed saw blade, it is actually displaced by the blade. In some cases, residual heat is required in the extrudate to allow material displacement without fracturing. Many different shapes of blades are available to allow chopping or slicing for different materials. Different edge geometry and materials also are available to enhance cut quality and blade life. Also,

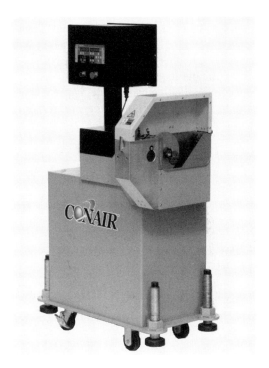

Figure 9 Rotary knife cutter.

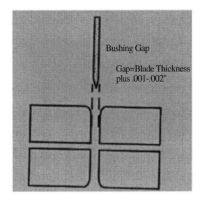

Figure 10 Bushing gap schematic.

materials such as isopropyl alcohol may be used to minimize material sticking to the blade or bushings, further minimizing particulate generation. With rotary knife cutters, cuts may be activated either in on-demand or continuous modes, depending on the cycles per minute required.

Rotary knife cutters can operate in an on-demand mode, where a single cut cycle occurs per input. Up to 350 cuts/min or more are common. Alternatively, rotary knife cutters can operate in a continuous, or flywheel, mode. One or more blades are attached to a speed-controlled wheel that rotates at a constant speed. The diameter of the wheel directly relates to the blade velocity, and is typically sized to minimize interruption. In this mode, up to 12,000 cuts/min or more are possible.

Typically, brushless servomotors are used to cycle the blade in rotary motion. For pharmaceutical cutting applications, clutch brakes should not be considered due to particulate generation from the clutch brake unit itself. Servomotors are fully sealed units and are supplied with stainless steel shafting, and offer clean and quiet operation. For higher cutting torque requirements, an in-line planetary reducer may be C-faced to the servomotor to increase cutting torque and motor efficiency.

On-demand cutting modes use input devices such as timers, encoder counters, and photosensors to activate a cutting cycle. Upon input, a single revolution of the blade is made by cutting the extrudate in the middle of the revolution. For continuous cutting modes, the speed of the wheel follows the puller speed at a set ratio. In this matter, if the puller speed is changed, the cut length remains constant.

The rotary knife cutter, which is placed downstream of the puller, requires a consistent delivery of material. The puller speed consistency directly effects the cut-to-length tolerances. The distance between the puller and the cutter will depend on the rigidity of the material being extruded (see Figure 11). If the material is very flexible, a very short distance is necessary to facilitate consistent feeding. Rigid materials will require distance between the puller and the cutter, allowing deflection during the stoppage created by the blade passing through the part.

Regardless of whether the cutter is on-demand or flywheel, the face of the bushings, between which the cutting blade passes, must be properly maintained to minimize particulate generation caused by scratches or wear. Think of the surfaces of the blade and the bushings as a pair of scissors, which must be sharp to facilitate shearing against one another. This allows a minimal area for the material to be pulled down between the surfaces, causing tearing and/or particulate generation.

Figure 11 Puller/cutter combination.

In many cases, a form of blade wipe is used to wash the blade between cuts and to lubricate it to minimize "sticking." A blade wipe may consist of a piece of absorbable material sandwiched between stainless steel plates, with a controlled drip system wetting it. The guard itself may include a tray of water or alcohol. During the cutting cycle, the blade simply passes through it, wetting the surface. This is the simplest and often the best method of lubricating a blade to minimize the sticking of the material to the blade and its transferring to the next cut.

Discharge conveyors are commonly used after the extrudate is cut to both support the part while the cutting cycle is taking place and to remove the cut part in a controlled manner. It is typical for the speed of this conveyor to be 10–15% faster than the line speed to create a part separation. In many cases, an automatic ejection system is incorporated into the conveyor to eject or divert parts off the side for batches or boxing. A delay and duration signal is

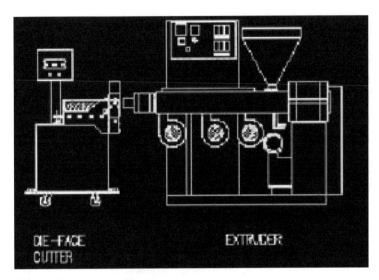

Figure 12 Extruder die.

generated from the cut signal to coordinate the position of the part to be ejected. It is also possible to use an in-line gauging equipment to signal the ejection system to automatically separate out-of-specification parts.

In some pharmaceutical extrusion applications, the upstream cutter die is actually the extruder die (see Figure 12). This can be the case if the material has enough structure and density such that it can be die face-cut. This is the simplest extrusion method and requires no cooling or sizing after the die prior to cutting to length. A simple photo eye can be used to input length information to activate the cut cycle. A conveyor can be used to support the part during and after the cut, and to convey it to the next operation.

B. Coil Winding for Pharmaceutical Extrusions

Coil winders, which are typical in the plastic and rubber extrusion industry, can be easily modified and applied to pharmaceutical extrusion applications (see Figure 13). Winders are used to package extrusions, in-line on reels. These reels can be designed to allow a single row of the material or many rows, in which the extrusion is layered in rows from 50 to 750 mm wide. In some cases, the material itself becomes the package when a collapsible coiling head is used. The collapsible head offers support to the extrusion while layering the material.

Figure 13   Extrusion system with dual winder.

Once the desired length is attained, the material is strapped, or shrink wrap is applied to contain the material when removed from the collapsible head. For medical or pharmaceutical applications, the collapsible head is very typical, with the head itself being constructed of stainless steel.

These in-line winders must be capable of following the line speed of the extrusion. Either mechanical dancers or noncontact (ultrasonic) loop control devices are used to both monitor and control the speed of the winder to maintain a loop between the puller and the winder. In this way, minimum tension is exerted on the extruded material. Medical and pharmaceutical extrusions typically use the noncontact loop control typically with the addition of a stainless steel tray below the loop to ensure against contamination if the loop was to reach the floor.

Level wind units are available for even layering of the extruded material on the reels or collapsible heads. These units are easily set for different widths of material or shapes. Often guides are used to enhance the consistency of the layering of material to minimize crossover, or, in some cases, prevent the layers from touching one another.

Again, as with belt pullers, the drives of choice for medial or pharmaceutical applications are servodrives. With stainless steel shafting and totally sealed housings, winders are well suited for washdown. Of course, all exposed surfaces should be stainless steel and easily cleaned.

V. SUMMARY

Much of the current plastic extrusion downstream technology can be applied for pharmaceutical shape extrusion. The extrusion industry has always been quick to respond to new markets and will continue to work with users in the pharmaceutical industry to customize equipment as required for a *General Pharmaceutical product* installation. As with medical or food application, the processes and equipment that have historically been used for plastic extrusion are easily modified for pharmaceutical applications.

SUGGESTED READING

1. Bessemer RH. Advancements in Downstream Technology for Tubing, Rods, and Profiles. Medical/Pharmaceutical Extrusion Conference, Las Vegas, NV. December 5–7, 2001:345–382.
2. Martin C. In the mix: continuous compounding using twin screw extruders. Med Device Diagn Ind 2000.
3. Wagner JR Jr, Vlachopoulos J. The SPE Guide on Extrusion Technology and Troubleshooting. Brookfield, CT: The Society of Plastics Engineers, 2001.

12
Film, Sheet, and Laminates

Bert Elliott
American Leistritz Extruder Corporation, Somerville, New Jersey, USA

I. INTRODUCTION

One of the largest volume processes for extrusion within the plastics industry is the production of sheets and films. Many products are needed in sheet form, or at least start out as a sheet and undergo postprocessing afterward. Common examples are roll roofing, packaging films, flexible printing plates, filtration membranes, agricultural films and tarps, frozen dinner trays (thermoformed from sheet), building insulation, kitchen countertops, battery separators, various sheets for electronics applications, interior and exterior panels for automobiles, photographic and x-ray films, etc.

As with most extrusion equipment used in the pharmaceutical industry, the sheet and film processing machinery is almost all derived from units designed for processing plastics. Even though many of the pharmaceutical applications will be new, this has a real benefit in that the machine designs are well proven from 40 years or so of fine-tuning these devices for plastics processes. One problem that may arise, however, is that most pharmaceutical sheets and films will typically be much narrower in a transverse direction than high-volume plastic products. The most common application of sheet/film downstream equipment is for product packaging for medical and pharmaceutical products. Another application is for transdermal drug delivery systems where an active ingredient is intimately mixed with a carrier and applied to a substrate.

II. BASIC CONCEPTS

Most melt extrusion systems are generally thought of as having an imaginary separation line, immediately after the extruder (where the extrudate will emerge from the die). The first half of the system (the portion with the extruder) is considered the *upstream*, or *melt processing* section. The second half of the system is the *downstream*, or *takeoff* section. This chapter will deal primarily with the downstream section.

In designing or selecting a transdermal downstream system, it is best to start with the end product and work backward. The desired physical properties of the solid dosage form will dictate the configuration of the downstream system. Some important parameters are:

Extrusion throughput rate—This is typically expressed in kilograms per hour.

Finished product width—This is in the transverse direction, or across the web.

Product thickness—Products less than 0.005 in. are generally referred to as films, while products thicker than 0.005 in. are considered sheets.

Product tolerances—This will usually be expressed as a + or − figure, such as +0.002 or −0.002 in. For instance, if the product target thickness is 0.040 in., with a tolerance of +0.002 or −0.002 in., the extrudate can be anywhere in the range of 0.038–0.042 in. and still be acceptable. Tolerances should be arrived at very carefully. For instance, in the example above, merely changing the +−0.002 in. tolerance to +−0.001 in. could significantly increase the complexity and the cost of the machinery.

Sheet characteristics—Flexible, rigid, or semiflexible.

Final product of the system—Will the product be wound up on a core, or cut in lengths and stored flat?

Line speed—Line speed is usually expressed in feet per minute, and is determined by the extrusion throughput rate, die width and gap, material specific gravity, etc. Most products will only process with good stability within a certain range of line speed. Either too slow or too fast will be unstable and uncontrollable.

Web path—The web path is the route the material and any substrate will take from the die exit all the way to the end of the line. In designing this path, it is important to take into consideration the properties of the product as it moves through the path (e.g., as it cools, does the sheet become too stiff to make an "S" wrap around a roll?) Typically, it is best to work out the web path with the equipment manufacturer's input, as they may have experience with similar processes.

Type of drive system—Because the rolls directly determine the product thickness, speed control is critical. The rolls should be driven by a variable-speed motor, and have a closed-loop feedback drive giving +0.5% or −0.5% speed regulation or better. This can be accomplished with either a d.c. or a.c. motor drive. The latest a.c. flux vector drives have an advantage because of their tremendously wide speed range (1000:1 turndown ratio in some cases). The d.c. drives typically have a 10:1 or 20:1 turndown ratio. The motor will typically drive the roll through a gear reducer to achieve the proper roll speed range.

Gear pump integration—A gear pump can be used at the discharge of the extruder, to provide a more stable flow and adequate pressure to the die. This is particularly advantageous with a twin-screw extruder, whose outlet pressure may be inherently inconsistent. In almost every case, significantly tighter product dimensional tolerances will be attained when using a gear pump, as compared to a system without an outlet pump. A system using a gear pump requires a computer control system, as the pump inlet pressure

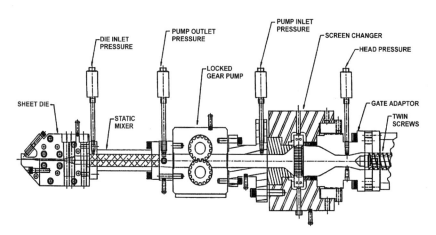

Figure 1 Gear pump front end.

must be controlled to a setpoint. The computer has a pressure control algorithm, which automatically modulates the feeder and screw revolutions per minute (rpm) setpoints to keep the pump inlet pressure under stable control (Figure 1) (1).

III. PROCESSES AND SUBPROCESSES

A. Processes for Making Film (Less than 0.005 in. Thick)

For making thin continuous films out of a polymer-based material, there are two main processes: blown film and cast film.

Blown Film

A blown film is named as such because the polymer is actually "blown" into a balloonlike shape using internal air pressure. This inflation is performed as the material exits a round die orifice shaped like an annular ring, and is accomplished under precisely controlled conditions in order to give a uniform film thickness. The extruded bubble is cooled by an outside stream of cool air supplied by an external blower. These are all performed continuously as the

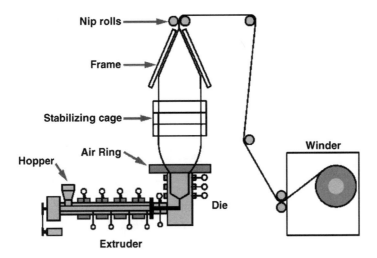

Figure 2   Blown film schematic.

bubble is being pulled upward vertically within a tower framework. At the top of the tower, the film is completely cooled, and is then progressively collapsed into a flat film "tube" (there are two layers at this point, having collapsed the two sides of the bubble flat), so it can be wound up into a roll (Figure 2).

A blown film can produce very strong thin films because it causes a biaxial orientation in the material as it is blown into the bubble shape. Orientation is a phenomenon that occurs within many polymer-like materials when they are stretched at a certain temperature, causing the long molecular chains to line up, or be oriented in a specific direction rather than just random directions. A common example is plastic garbage bags, which are only 0.0005 in. thick but are remarkably strong in all directions because they have been biaxially oriented.

Unfortunately, a blown film does not work with many materials—the critical parameter being the melt strength. Low-viscosity polymers, or polymers which do not have enough melt strength to maintain a film when blown, are not viable.

Cast Film

To make a cast film, a die is mated at the discharge end of the extruder to spread the melted material (or fluid) out into a thin "melt curtain" of uniform thickness. This type of die is referred to as a cast film die, and is commonly available in widths from 4 in. up to any width desired. Note that the finished film will generally be narrower than the width of the die opening because of "neckdown". Neckdown is caused because the roll speed will typically be adjusted so that it is actually pulling the film away from the die. For instance, a die with a 10-in.-wide opening may yield an 8.5-in. product width (see Figure 3).

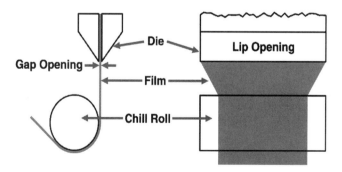

Figure 3 Cast film drawdown.

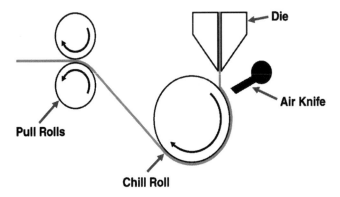

Figure 4   Cast film process.

A cast film is made by continuously "casting" a thin layer of material onto a rotating steel roll. The die is generally positioned either fully vertically (with the die lips pointing at the floor), or at a 45° angle. The roll surface is maintained at the desired temperature by pumping a liquid (typically water or oil) through the inside. Because the molten material is generally hot and solidifies onto the roll as it cools, the main function of the water is for cooling, although it may be used to preheat the roll during initial start-up (Figure 4).

An essential feature of a well-designed cast film takeoff unit is the height adjustment mechanism. When the extruder is started up or purged, there must be a provision to move the casting roll from underneath the die to allow for collection of the extrudate. This requires lowering the takeoff unit, and either putting a temporary "chute" in the way, or rolling the unit away from the extruder. When it is time to thread up the line and to produce the final film, the roll must be raised up into position close to the die lips. This mechanism may be a manual hand crank on a very small unit, or fully motorized on a larger machine. Production units typically use a motorized height adjustment.

Orienting

A cast film is typically oriented in the machine direction (Machine Direction Orientation, MDO) only. To biaxially orient cast film requires an immensely complex and expensive system called a tenter frame takeoff that heats and stretches the film after it has been cooled. The MDO is generally accomplished by having a separate motor drive on the pull roll station, so that the film can be stretched while moving along the web path. This will make a film that is very strong in the machine direction, but with normal strength in the transverse direction. Common examples of this in consumer-type goods are wrapping and strapping tapes (2).

Roll Design

The film casting roll is the key to the whole process. First of all, it must have the correct surface finish to impart the desired surface characteristics to the film. This may be a supersmooth mirror chrome for a glossy film, a medium mirror chrome, or a matte or other textured finish. Chrome plating has been used traditionally in the plastics industry because it offers excellent hardness and wear resistance. Rolls can also be fabricated from various grades of stainless steels, but are made mostly from 400 series stainless steel because of both weldability and hardenability.

The required roll diameter will be dictated from the throughput rate and the line speed. It is important to note that the roll is a heat exchanger, and therefore its capacity is proportional to its surface area. A casting roll that will handle 200 kg/hr will have to have a much larger diameter than one designed for 20 kg/hr. Also, the line speed must be considered because the material is leaving the die in a molten state and must be completely solidified by the time it leaves the roll. There will be a certain residence time required for the film to be in contact with the roll, and this, in conjunction with the line speed, will dictate the minimum diameter of the roll.

In order to remove heat from the roll surface and to maintain it continuously at the desired temperature, the internal geometry of the roll must be designed properly. A simple single shell roll will allow a liquid to be pumped through the interior of the roll, but the heat exchange is inefficient, and the roll surface temperature can vary considerably from end to end. This is because there is no predetermined path for the water to take—it is only flowing in a random pattern, and at a slow velocity. This type of design is generally not deemed appropriate for pharmaceutical products.

The optimum roll construction is called "double-shell, spiral baffled" roll. A film casting roll has a thin outer shell to optimize heat transfer capabilities to cool the product. Double shell refers to the second roll shell within the inside of the main roll shell, with a gap of typically 10 mm between the two shells. Spiral baffled refers to the fact that within this annular space, there is a helical barrier welded in place. This forces the coolant to flow only within the spiral channel, keeping it along the inner wall of the main roll shell. Because this channel has a small cross-sectional area, the water flows at a higher velocity, making convection more efficient (Figure 5).

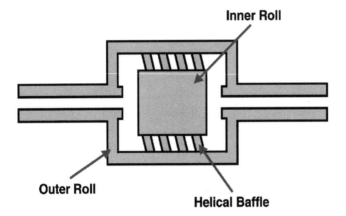

Figure 5   Double-shell, spiral baffled roll.

Pull Rolls and Slitter

After the film leaves the main casting roll, the rest of the process is very straightforward. The film generally goes through a set of rubber-covered nip rolls, and from there to a winder. Note that film made with most materials will have an "edge bead," meaning that the outer edges of the film will be thicker than the middle. For some materials and purposes, the effect is not noticeable and can be used as is with the edge bead on.

If the edges must be slit off, the slitter station would typically be located after the nip rolls, and prior to the winder. Depending on the type of film (and type of winder), this may require another set of nip rolls after the slitter. Slitters for films can usually use simple fixed razor knives, unless there is a laminated substrate layer that is difficult to slit, in which case a motorized rotary blade system may be required.

Laminations

A film casting unit can also be outfitted with one (1) or more unwind stations to cast the film onto a substrate. The position of the unwind must be chosen by taking into account the proposed web path to bring the substrate into the casting roll at the correct place to allow adhesion. In general, there are only two possibilities, with the unwind shaft being either above or below the casting roll. Sometimes a front rubber nip roll is necessary to apply nip pressure to the substrate/film to gain proper adhesion strength (Figure 6).

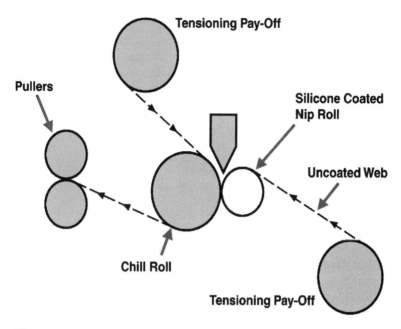

Figure 6 Typical lamination line.

Unwind designs can range in sophistication depending on the tension control accuracy required. To determine the tension needed, the substrate properties must be evaluated carefully. If it is a strong, thick substrate that is not stretchy, the equipment will work with a fairly wide range of tensions. On the other hand, if the substrate web is thin and weak, or will wrinkle easily, then it is important to have a better control of tension to successfully laminate.

An unwind shaft must have a brake mechanism to apply drag to the shaft to produce tension in the substrate web. The simplest type of brake is mechanical, using a friction drum or disc either applied with a hand adjuster, or pneumatically. A smoother braking drag can be attained using a magnetic particle brake, where the drag is varied by changing the voltage to the brake. With any system, it is important to determine if a manually set drag is acceptable, as this type will not automatically "taper" or compensate for the change in substrate roll diameter. This is required because as the material is unwound from the payoff, the roll diameter gets smaller and smaller—changing the actual web tension.

The optimum arrangement is where there is an automatic taper compensation for change in a roll diameter. This is usually done by having the web make an approximately 90° turn over an idler roll that is supported on load cells, so it is always measuring the force on the idler roller. This force signal is fed back to a controller, which has a closed loop to the brake control. In this way, the operator enters a tension setpoint, and the control modulates the brake to maintain this tension, regardless of substrate roll diameter.

Another hardware issue to be considered is how to affix the substrate roll onto the unwind shaft. There are various methods, depending on how often there is a need to change the rolls, and how quickly this operation must be done. Most substrates are supplied on 3-in. ID plastic cores. Cone collars are a simple, tried-and-true method of holding the core, but are not convenient if the core changes are often, or quick while running. Air chucks are a better choice for production applications. An air chuck uses compressed air to inflate a rubber bladder within the shaft, which expands aluminum "leaves" or "buttons" to grab the core. With an air chuck, the operator merely presses a switch to either release or grab the core, which permits easy, rapid changeovers.

A final bit of hardware that may be needed is a transverse adjustment for each unwind. This is necessary if the edge alignment of your product is important. The transverse adjustment is usually accomplished by having the unwind mounted to a carriage on linear bearings, with a simple threaded rod and manual handwheel for adjustment back and forth.

Air Knives and Edge Pinners

Sometimes a condition can develop on the main casting roll immediately downstream of the die, where a layer of air gets continuously trapped in between the film and the roll surface. This entrapped air will cause a reduction in heat transfer, and the finish of the film will suffer as a result. One device to alleviate this problem is an air knife. An air knife puts out a "curtain" of cool air about the same width as the film, which is directed against the film to keep it in contact with the roll. On a small unit, air knives are generally compressed air operated. In addition to keeping the film pressed against the roll, the air knife also provides some auxiliary cooling of the opposite side of the film.

Edge pinners are similar devices, but instead of a curtain, these devices emit a "point" jet of compressed air. There are typically two jets, one for each edge. The reason edge pinners are used is that some materials may have heavy flow at the outer ends of the die, producing a thick edge bead. This edge bead can then lift the film off the roll surface or cause alternating "scalloping" on the edge. The jet of air is directed to pin the edge down against the roll surface, thus improving the overall uniformity of the film.

Film Winders

As with unwinds, there are many different types of winders. Probably the single most important concern is the required winding tension. Some films require very delicate tension held within a tight tolerance range; otherwise, wrinkles will result. For these types of films, an automatic taper tension control system is necessary.

Other important parameters are:

the diameter of the wound roll;
the weight of the final roll;
whether a cantilevered design is possible;
the winding feet per minute; and
the method for core changeovers.

Spreader Rolls

Thin films of many materials have a tendency to wrinkle and to "bunch together" when under tension. The film can even "fold over" in places, making it impossible to wind into a uniform roll. The solution to most of these problems is to use a spreader roll. A spreader roll is designed to continuously pull the film from the center out to both edges. There are many different designs of these rolls from various manufacturers. Some are rubber-covered, and are warped into a curve (a "banana roll"), and others may be aluminum or stainless steel with helical grooves cut into the roll.

The other factor that must be decided is where to put the spreader roll. Spreader rolls are typically used just prior to the winder, but others may be placed in the web path depending on the intended process.

B. Processes for Making Sheet

Sheet Takeoff Machines

Many of the same comments noted above for cast film also apply to sheet extrusion. Sheet extrusion also uses a flat-type die, and water or oil cooled rolls to form the product. In general, a sheet equipment is of a heavier construction than film machinery, as the roll nip forces are much higher. Sheet also tends to run at much slower line speeds because the product is thicker. Many sheet products run in the range of 1–30 ft/min.

Sheet takeoff units are made in several different configurations, with the main ones being Vertical stack, 45° Canted stack, and Horizontal stack (referring to the positions of the main cooling rolls). Vertical and Canted stacks can be either "upstacks" or "downstacks," which refer to which direction the web is going to move away from the die. Some products will run equally well on any of these configurations, but sometimes a material will be much easier to run on a particular type of machine. For instance, a low-viscosity material may not have enough melt strength to work well with a vertical stack because the melt curtain will sag in between the die and the primary roll nip. For this type of material, either a canted or horizontal stack will work better, as the material can fall by gravity right into the roll nip (Figures 7 and 8).

A major design criterion that must be answered when specifying a sheet takeoff is, "What is the required nip force of the rolls to "squeeze" the materials to make the product?" This is defined in pounds per linear inch

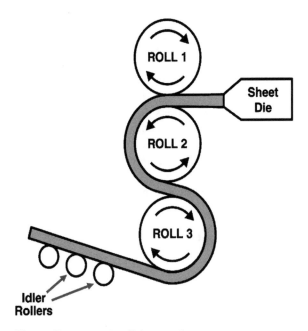

Figure 7   Sheet takeoff downstack.

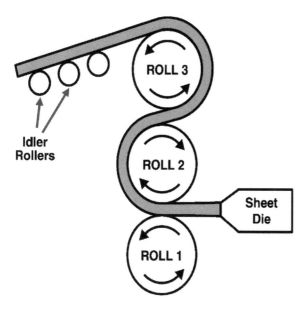

Figure 8 Sheet takeoff upstack.

(PLI). The design of the rolls, bearings, and frame of the takeoff machine depends greatly on the rated PLI. A light-duty machine would typically be capable of 150–200 PLI. A medium-duty unit would be typically rated for 350–400 PLI, and a high PLI machine would be in the 700–1000 range (Figure 9).

As a general rule, low melt viscosity materials will not need much PLI force, while tougher, higher molecular weight materials will need more. If the machine does not have enough PLI, the rolls will be pushed open by the melted material, with the end result being inconsistent gauge control (because the roll arms are not being held firmly against the gap stops). The best approach is to test the materials on an existing takeoff unit to determine the required PLI for the application.

Sheet Dies

A sheet die looks similar to a cast film die from the outside. Both take a round melt input and distribute it into a melt curtain of uniform thickness and desired width. The sheet die will usually need a wider range of gap openings, however, as the product may range from 0.010 to 0.125 in. or more. Sheet

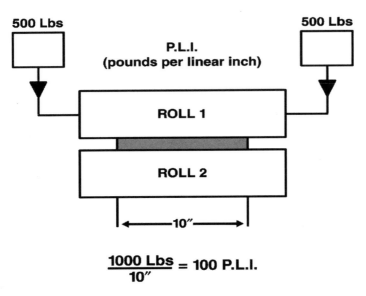

Figure 9 Example PLI.

dies are offered in light-duty and heavy-duty designs, with the choice depending on the internal pressure expected.

Lip gap adjustment mechanisms for almost all modern dies are of the flex lip type (Figure 10). Flex lip means that a row of bolts is used to adjust the lip by actually flexing a thin section of steel within the die. One feature that is different from a cast film die is that a sheet die very often has a choker bar, particularly to run thicker products. A choker bar is a moveable dam inside the preland area, which can be adjusted to reduce the gap in the preland. This serves to increase the pressure within the distribution manifold, improving the uniformity of the melt curtain.

Flex lip mechanisms are offered in a few different designs also. The simplest type is called "push only" because the adjuster bolts can just push on the lip to close it down to a smaller gap. A more complex lip adjustment design is "push–pull," meaning that the adjuster bolt can both push and pull the lip.

The most sophisticated and expensive lip adjustment method is an Automatic die (or Auto die, for short). This is a system using a noncontact thickness scanner located in the downstream web path, near the pull rolls, where the sheet is fully cooled. The scanner traverses the full width of the sheet back and forth continuously while running, and takes very precise

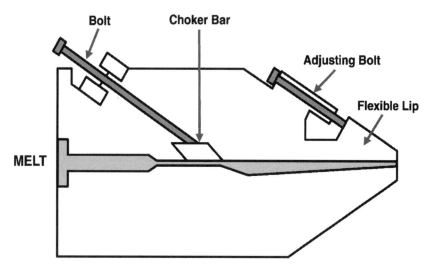

Figure 10   Flexible lip sheeting die.

thickness measurements. Most scanners today use a radioactive material that emits beta or gamma rays. These thickness measurements are collected in a computer, which correlates them to the transverse position of the measuring head. The software for these systems is very sophisticated, providing a multitude of real-time Statistical Process Control data in addition to the thickness profile.

The bolt adjustment controls automatically adjust the die lip to correct it. This is generally done via special heated bolts in place of the normal lip adjustment bolts (3).

Sheet Slitters

Sheets generally always have an edge bead on both sides, which often must be slit off. Thinner sheet gauges in flexible materials may use a simple razor slitter, identical to the unit described above for film. Heavier gauge sheets or tough materials will usually need a powered shear or score cut slitter.

Web Guiding

Web guiding is a term used to describe a feature to either manually or automatically guide the sheet to keep it centered on the rolls as it progresses along the web path. This is necessary with a web that tends to walk off to one side. Web guiding may not be necessary with a relatively short web path from die to winder. As this path gets longer with more and more stations in between, web guiding may become necessary. The most sophisticated systems use an edge sensor (typically a photo eye), which picks up the direction the edge is drifting. This is fed back to a motor-controlled pivoting idler roll carriage that can skew the sheet one way or the other.

Roll Temperature Control

Rolls are maintained at a desired temperature by circulating a liquid through the inside of the shell. This liquid is generally either water with some additives such as ethylene glycol or oil. Water may be used with pressurization up to about 100°C; above this temperature, oil must be used. Oil is less desirable for a few reasons:

1. The thermal conductivity of oil is only about 60% that of water, meaning heat transfer is less efficient.
2. Oil is very messy when there is a leak, or a rotary union must be taken apart.
3. Most heat transfer oils degrade over time at elevated temperatures, so that the oil must be changed at certain intervals.
4. Oil pumps and temperature control units generally require more maintenance than water units.

Whether water or oil is used, temperature control units are required for each independent "zone" of control. With a roll stack with three rolls, it is possible to operate all three from a common temperature control unit, but this results in the same temperature for all three rolls, which may be detrimental to the process. A more flexible setup would be to have an independent unit for each roll, allowing individual temperature set points for each roll (Figure 11).

Each unit will have a heat exchanger for cooling, so that even if it is an oil unit, a source and a return of cooling water must be supplied.

Sheet Processing Tips

Like many extrusion processes, there are many nuances that result in a successful operation. An issue that can have a significant effect on the sheet properties is the degree of "drawdown." Drawdown is simply the linear speed of the takeoff rolls relative to the linear speed of the melt curtain exiting the die. If these speeds were exactly matched, the sheet would be taken away at the same speed as the material coming out of the

- **ROLL 1 contacts the sheet only along its tangent and conducts little heat.**

- **ROLL 2 absorbs heat for one half of its surface and conducts the most heat.**

- **ROLL 3 absorbs heat for one half of its surface but after ROLL 2 has reduced the total heat energy in the sheet.**

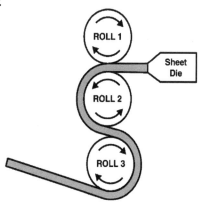

Figure 11 Differential temperature control.

die. This can be done with some materials, but very often this is not the most stable way to run. Because the melt curtain has no "melt tension" on it, the sheet may wander and not cool in a uniform way, resulting in poor gauge control. Because the rolls are just "kissing" the sheet, surface contact may be lost from time to time, which also may cause an uneven surface texture.

Many materials process best with the takeoff rolls running a little faster than the material exiting the die. This provides a definite drawdown to the product, meaning it will be both thinner and narrower (necked down) than the die lip dimensions. So this has to be taken into consideration when setting up the die lip gap. For instance, if a 0.050-in. product is desired, the die lips may be set to 0.065 in., with the product drawdown using roll speed. Note that drawdown can create different physical properties in the sheet for machine direction versus transverse direction. The more the product is drawn, the greater this effect. Drawdown can be used to optimize the material properties. For instance, some rigid materials—when run with excessive drawdown—will be very strong in one direction, and easily crack if flexed in the other direction.

It is important to note that the roll gap is not necessarily set to the exact dimension as the final product gauge. This again is highly material dependent. Because the sheet is still cooling inside as it leaves the primary nip, it may either continue to swell or expand, or it may contract upon subsequent cooling. This even depends somewhat on what meters per minute speed range is operated. The correct gap is generally arrived at by performing test runs with known roll gap settings, and by checking the gauge of the cooled sheet.

The opposite condition of drawdown is to run what is called a "rolling bank" (Figure 12). This effect is created when the rolls are run at a slightly slower linear speed than the melt curtain is exiting the die. This causes a "pool" of molten material to build up continuously in the primary nip gap. Because of the surface friction of the rolls, this pool tends to roll over itself continuously, hence the term rolling bank. The rolling bank is a pool of excess material, so one benefit is that it ensures that the nip gap is 100% filled, making for good contact between the rolls and the sheet.

Running a rolling bank can be a delicate balancing act. If the pool is too large, it will cause excessive forces on the rolls, which can actually cause damage to the steel face of the roll. A large rolling bank may also induce residual stresses into the sheet product, or cause visible arc lines in the sheet examined under a light. The decision to process with a rolling bank is another decision that is material-dependent. Some materials with a certain melt viscosity and elasticity run very well this way; others do not (4).

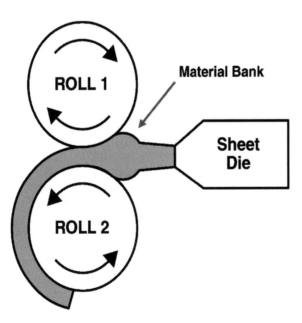

Figure 12 Rolling bank.

IV. GMP REQUIREMENTS

As mentioned in Sec. I, almost all of this type of equipment are derived from units designed for the plastics industry. For each machine built for a pharmaceutical customer, the manufacturer may be building 100 units for plastics processors. Various machinery builders have different degrees of experience with pharmaceutical practices. Some are very familiar with the requirements, while others are not. The safest approach when purchasing a GMP machine system is to generate a detailed document of specifications for both mechanical and electrical components within the system (5).

The GMP considerations can have a significant effect on the cost of the equipment. For instance, one way to build the equipment is with material contact surfaces made of stainless steel, and supporting frames and other noncontact parts of painted carbon steel. This is very cost-effective, and may meet your requirements. Another variation is to allow painted carbon steel below the process height, and to require stainless steel above the process height (to prevent paint chips from accidentally falling into the material). Some GMP systems require everything to be fabricated from stainless steel, including machine bases and framework. This will involve significantly higher cost.

The level of finish expected on the rolls should also be specified, and in quantitative terms. It is far better to specify that the "rolls will have a finish of 0.8–1.5 μin." than to merely say "mirror finish."

V. CONCLUSION

The technology to continuously produce a sheet or film is well proven, with literally thousands of installations in North America alone. Currently, the challenge is applying this well-proven technology to the pharmaceutical industry. Generally, this does not relate to design and process issues, but rather to "downsizing" the equipment and configuring the machinery for a GMP environment.

REFERENCES

1. Martin C. Using Twin Screw Extruders to Perform Compounding with Direct Sheet Extrusion of TPO Formulations. TPO RETEC: Troy, MI, Sept. 20, 1999.
2. Christie A. Machine Direction Orientation of Thin Films: Properties and Processes. Film and Sheet Conference 97, Somerset, NJ, Dec. 9, 1997.

3. Cloeren PF. Getting The Most Out of Your Die System Investment. Film and Sheet Conference 97, Somerset, NJ, Dec. 9, 1997.
4. Lamont PR. Equipment and Processing Considerations for Thin Gauge PP Sheet. Film and Sheet Conference 97, Somerset, NJ, Dec. 9, 1997.
5. Martin C. Continuous Mixing/Devolatizing Via Twin Screw Extruders for Drug Delivery Systems. Interphex Conference, Philadelphia, PA, March 20, 2001.

13

Melt-Extruded
Molecular Dispersions

Jörg Breitenbach and Markus Mägerlein
*Knoll Soliqs, Abbott GmbH and Company,
Ludwigshafen, Germany*

I. INTRODUCTION

When is an active molecule a good drug? Companies are developing new screening tools to help them select druglike candidates early in the drug discovery process. Determining an active compound's "drugability" is a bottleneck that can involve hundreds of tests to determine bioavailability and toxicity (alone and in combination with other drugs); this comprises a set of characteristics known as administration distribution metabolism excretion toxicology (ADMET). However, one of these aspects, namely the improvement of the solubility, dissolution rate, and absorption properties of poorly water-soluble drugs, remains a challenging aspect in the development of pharmaceutical products.

One approach for drugs whose gastrointestinal (GI) absorption is limited by dissolution, especially drugs with poor water solubility, is the reduction of particle size, which generally increases the rate of dissolution (Noyes–Whitney) and therefore of absorption and/or total bioavailability. This was dramatically illustrated for griseofulvin (1) and spironolactone: the dose of each of these drugs was halved by reducing particle size.

II. SOLID DISPERSIONS

Sekiguchi and Obi demonstrated the unique approach of solid dispersions to reduce particle size and to increase rates of dissolution. They prepared the eutectic mixture of sulphathiazole with a physiologically inert and easily soluble carrier (2). They were also the first to report the melting or fusion method. A physical mixture of an active agent and a water-soluble carrier is heated until it is melted. The melt is solidified rapidly in an ice bath under rigorous stirring, is pulverized, and then is sieved. Rapid congealing is desirable because it results in the supersaturation of the drug as a result of entrapment of solute molecules in the solvent matrix by instantaneous solidification.

In 1974, in his article "Dissolution of Solid Dispersion Systems," Hajratwala (3) stated that, "Solid dispersion of drugs is a relatively new field of pharmaceutical technique and its principles play an important role in increasing dissolution, absorption, and therapeutic efficacy of drugs." When we look at the market of pharmaceutical products today, solid dispersion systems are still found to be neglected. Only a few ever made it to the market, including griseofulvin–polyethyleneglycol–dispersion (Gris-PEG™ marketed by Wander); Cesamet™, a nabilone polyvinylpyrrolidone (PVP) preparation (marketed by Lilly) (4); as well as a formulation of Troglitazone (Rezulin™) marketed by Parke-Davis (5), which had to be withdrawn from the market due to toxicology-related issues of the drug active. Problems limiting the commercial application of solid dispersions involve the method of preparation, the reproducibility of physicochemical properties, the formulation into dosage forms, the scale-up of manufacturing processes, and the physical and chemical stability of the drug and the vehicle. Solid dispersions of drugs were often produced by melt or solvent evaporation methods. The materials, which were usually semisolid and waxy in nature, were hardened by cooling to very low temperatures. They were then pulverized, sieved, mixed with relatively large amounts of excipients, and encapsulated into hard gelatin capsules or compressed into tablets. These operations were difficult to scale up for the manufacture of dosage forms (6).

Melt extrusion is a significant step forward to cover the technology-related issues, and makes the solid dispersion approach a more interesting option.

Basically, solid dispersion systems can be divided into six different morphologies: simple eutectic mixtures, solid solutions, glass solutions or glass suspensions, amorphous precipitations in a crystalline carrier, compound and complex formation, and combinations of the previous five types (7). However, often, a solid dispersion does not entirely belong to any of the groups, but is rather a combination of groups, and therefore the desired effect such as an enhanced dissolution rate may also be attributed to different mechanisms (Figure 1).

A matter of uncertainty rarely addressed is the analytical differentiation of amorphous embeddings in crystalline carriers compared to a true solid solution, where the drug is molecularly dispersed in the carrier. The same holds for the amorphous embedding in a glassy carrier compared to a glass solution. We therefore propose the term molecular solid dispersion in order to differentiate from amorphous embeddings. The molecular solid dispersion summarizes solid and glass solutions. It is defined as the molecular embedding in a carrier that is a glass or crystalline matrix.

In spite of the "activated state" of the drug substance in the molecular solid dispersion, the state of the drug cannot be compared with the mere metastable amorphous form. The molecularly dispersed state of the drug substance is stabilized by the polymer matrix. Two major factors that stabilize molecular solid dispersions are intermolecular interactions between the drug and the carrier, and the viscosity (8) of the carrier. An evidence for solid-state and liquid-state interactions in a furosemide–PVP solid dispersion was demonstrated by Doherty and York using spectral subtraction procedures. The proposed hydrogen bond–furosemide–PVP interaction may account for the formation and the stabilization of the molecular solid dispersion (9).

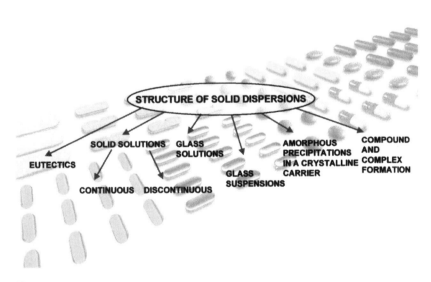

Figure 1 Structure of solid dispersions.

Glass transition temperature has long been seen as the predominant factor governing the physical stability of solid dispersions. As a general concept for stabilizing solid dispersions, the enhancement of the glass transition temperature by adding excipients with higher glass transition temperatures has been promoted. However, this concept only holds for true molecular solid dispersions. Within amorphous solid dispersions, the mobility of the active is already given in the amorphous phase of the active itself, possibly lending to recrystallization processes. In the meantime, investigations have shown that direct linear correlations between glass transition temperature and recrystallization tendency are rarely given, and that the solubilizing and stabilizing effects of the carrier systems intermolecular interactions are often of far greater importance for physicochemical stability (10). Especially with PVP as a reliable hydrogen bond acceptor, complex formation might be extremely important (11).

Nevertheless, the formulation as a supersaturated molecular solid dispersion or as an amorphous embedding bears the intrinsical risk of reverting to the more stable crystalline form. In order to address this question, the importance of timescales of molecular motion in amorphous systems must be considered (12).

We performed studies with respect to the solubility of the drug and the matrix viscosity, as well as glass transition to investigate these important factors (13).

The contribution of different mechanisms enhancing drug release from solid dispersions with particle size reduction following the Noyes–Whitney equation has also been discussed (14). However, the Noyes–Whitney equation does not encompass the molecular solution, as in this case, the matrix having dissolved the drug molecule is the predominant factor.

The mechanisms of increased dissolution rate have been described recently by Corrigan (15). The main reasons are as follows:

1. In the case of glass, solid solutions, and amorphous dispersions, particle size is reduced to a minimum level. This can result in an enhanced dissolution rate due to both an increase in the surface area and solubilization.

2. The carrier material, as it dissolves, may have a solubilization effect on the drug.

3. The carrier material may also have an enhancing effect on the wettability and the dispersibility of the drug in the dissolution media. This should retard any agglomeration or aggregation of the particles, which can slow down the dissolution process.

4. The formation of metastable dispersions that have a greater solubility would result in faster dissolution rates.

Solid dispersion systems for drugs have been discussed extensively with particular respect to the methods of preparation, the need to optimize the drug/carrier ratios, and the need to maximize dissolution and absorption rates. Among the suitable polymeric excipients, PVP (16) or its copolymers (17), poly(ethylene-co-vinylacetate) (18), poly(ethylene-oxide), cellulose ethers (19), acrylate (20), and other matrices have been applied. The basic prerequisite for their use in melt extrusion is the thermoplasticity of the polymers. The thermoplasticity may be influenced by plasticizers, often a property of the active ingredient itself. Also unmeltable carriers have been used. Based on the concept of preparing a powder from a glassy drug dispersed in an inert carrier by melting, Nakagami (21) produced amorphous drug–silica mixtures.

Solid dispersions of drugs with poor solubility revealed remarkably higher bioavailability (22). Obvious potential advantages are lower drug dose and reduced interpatient variability. Less drug material may be applied, and potentially a lower degree of variability is an obvious advantage.

III. MELT EXTRUSION TECHNOLOGY

A. Background

The processing of melts in order to obtain solid dispersions is well known (23), and the essential advantage of a melt process in this domain is its solvent-free formation of solid dispersions. Pioneering work in the field employing the melt extrusion process as a manufacturing tool was performed by Doelker et al. (24).

Whereas the extrusion of wet masses is a standard technology in the field of pharmaceutical production, melt extrusion so far has been mainly used in polymer engineering (25). Melt extrusion for the manufacture of pellets (27,28) had revealed the potential for a controlled release of polymer-embedded drugs and the limitations of the technology. Also a related process, injection molding, has been used to manufacture solid dispersions (29) and has been extended into a drug delivery technology (30).

The properties of polyethylene oxide (PEO) as a drug carrier in melt extrusion, with the aim to obtain chlorpheniramine maleate matrix tablets, was examined by Zhang and McGinity (31). They also examined the use of melt extrusion for effervescent tablets (32). Sarraf et al. (33) recently demonstrated that a 200-μm-thick layer on the extrudate surface can be introduced while extruding and can thereby alter the release characteristics.

The breakthrough in the field of melt extrusion was achieved by making a great variety of carrier systems feasible using, e.g., polymers and additional excipients in a comparatively simple integrated technological system.

A floating dosage form composed of nicardipine hydrochloride and hydroxypropylmethylcellulose acetate succinate was prepared using a twin-screw extruder. By adjusting the position of the high-pressure screw elements in the immediate vicinity of the die outlet, and by controlling the barrel temperature, a puffed dosage form with very small and uniform pores was obtained. It was shown that the puffed dosage form, consisting of enteric polymer prepared using the twin-screw extruder, was very useful as a floating dosage form that was retained for a long period in the stomach (34).

17-Estradiol hemihydrate as a poorly water-soluble drug improved in its solubility and dissolution rate by melt extrusion. Different compositions of excipients such as PEG 6000, PVP, or a vinylpyrrolidone–vinylacetate copolymer were used as polymers and Sucroester™ WE15 or Gelucire™ 44/14 as additive. The solid dispersions resulted in a significant increase in dissolution rate when compared to the pure drug or to the physical mixtures. A 30-fold increase in dissolution rate was obtained for a formulation containing 10% 17-estradiol, 50% PVP, and 40% Gelucire 44/14. The solid dispersions were then processed into tablets. The improvement in the dissolution behavior was also maintained with the tablets (35).

The patent literature provides further examples. Underlining the economic importance of the melt extrusion process, the use of different polymers was examined, revealing that hydroxypropyl cellulose may be a better water-soluble polymer compared to PVP for poorly soluble drugs. X-ray powder diffraction (XRPD) studies suggested that the drug substance mostly existed in the crystalline state, truly representing a crystalline solid dispersion. Extrusion as a way of manufacture is mentioned and a variety of examples using melt processes are given (36).

Further, the extrusion process has been applied in order to combine a sparingly water-soluble drug compound, cyclodextrin, and a physiologically tolerable water-soluble organic polymer (37).

Melt extrusion with pharmaceutical actives having a low melting point has also been described. Under such conditions, the extrusion process may serve as an agglomeration step (38).

Also the influence of plasticizers and drugs on the physicomechanical properties of hydroxypropylcellulose (HPC) films prepared by melt extrusion has been investigated (39).

The correlation between T_g and the concentration of the individual components of a glass solution is given by the Gordon–Taylor equation. However, the Gordon–Taylor approach assumes that ideal mixing and negative deviations may be related to interactions especially hydrogen bonding (40).

Amorphous polysaccharide matrices determining the release rate of melt-extruded dispersions have been examined (41).

Predicting when recrystallization will occur using T_g has been suggested as a useful way of estimating the amorphous physical state. Matsumoto and Zografi demonstrated that the major basis for the crystal inhibition of indomethacin at 30°C at the 5% wt/wt polymer level in molecular dispersions is not related to polymer molecular weight and to the glass transition temperature, but is more likely to be related to the ability to hydrogen bond with indomethacin and to inhibit the formation of carboxylic acid dimers that are required for the nucleation and growth of the crystal form of indomethacin (42). Forster et al. melt-extruded indomethacin, lacidipine, and tolbutamide with PVP. Indomethacin samples remained amorphous on storage as proven by XRPD and temperature-modulated differential scanning calorimetry (TMDSC). For lacidipine and tolbutamide, low crystallinity levels were detected, which (at estimated 7% crystallinity) decreased the dissolution rate significantly. The presence of significant hydrogen bonding (as indicated by the deviation of the experimental temperature from the theoretical glass transition temperature, T_g, values) may decrease the uptake of moisture and restrict the plasticization of the matrix. However, lacidipine recrystallized even when T_g was well above the storage temperature. In their conclusion, the authors talk about the amorphous drug in the matrix, which is not in line with the term glass solution (43).

In general, the melt extrusion process is a solvent-free manufacturing process. It offers the chance to embed drugs in the molecular state and is capable of handling actives of different particle sizes as well as amorphous solids, thus providing solutions to polymorphism and pseudopolymorphism-related problems. Therefore it represents a process that is specifically tailored to handle these questions.

The effect of the molecular weight of the matrix polymers on the dissolution profile has been described with special respect to solid dispersions (44). The mechanism of dissolution of molten matrices strongly depends on the choice and the combination of polymeric excipients. Another factor predetermining the release rate is the status of the drug, which can again be present either as crystalline or amorphous particles, or actually dissolved in the polymeric matrix. In the latter case, the active is an intrinsical part of the matrix, thus influencing its wettability and release characteristics. Information on the interaction between water and polymers is indispensable in manufacturing a solid dispersion of a drug by hot melt extrusion because this interaction affects various properties of the water–polymer mixtures, such as their viscoelastic properties (54). In addition, molecular weight represents a tool to determine the process temperature and the energy input via shear forces (45).

Today, melt extrusion has made significant progress in developing from a process technology to a drug delivery technology. The intriguing combination of molecular solid dispersions and controlled release (46) can be of advantage when it comes to decreased absorption because the poor solubility of the active limits the availability of the drug in the lower GI tract. Moreover, the molecular solid dispersion concept has been reported to also offer benefits due to reduced gastric irritation (47). Additional processing advantages include circumventing the problem of polymorphic forms with different solubilities, the possibility to start from different particle sizes, and the possible reduction of tablet size. Examples have shown that more favorable polymorphic forms may be stabilized upon incorporation into PVP (48).

B. Process

The extruder—typically with a corotating twin-screw configuration—is one core element in the technological system (Figure 2). This equipment consists

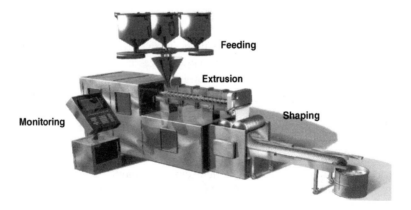

Figure 2   Schematic presentation of the melt extrusion process.

of a hopper, barrels, screws, kneaders, and die. Feed screws and a kneading paddle are incorporated into the two screws and the screws rotate equiaxially, accurately engaged within the barrels. In addition, the screw elements and barrels are of a structure whereby the type and the arrangement can be changed, according to the requirements (e.g., viscosity of the plastified mass or sensitivity of the active) (49). Its operation is flexible, easily controlled, and can be well documented.

The temperatures of all the barrels are independent and can be accurately controlled from low temperatures (30°C) to high temperatures (250°C) (50). During the passage through the extruder, the mass is heated either by external heating and/or shear forces, and the polymer matrix is thereby plasticized to incorporate the drug material. If solid melt extrusion dispersions are prepared using the melt extrusion process, no solvents may be applied and the drug dissolves in the polymeric matrix. In such case, the matrix can be regarded as a highly viscous solvent.

The second core element of the integrated technological system is the device to shape on-line the molten strand leaving the extruder. Basically, there are two different ways of doing this:

1. Calendering, in which the molten strand is forced between two calender rollers, thus producing films, flakes, or sheets, which may already contain single tablet cores; and
2. Pellet forming, which may, for example, be a rotating knife cutting spaghetti-like extruded strands.

Overall, the residence time in the extruder is rather short (approximately 2 min) as the process conveys the extruded mass continuously throughout the extrusion channel, thus avoiding heat stress on the active and the respective excipients in the formulation. In addition, oxygen and moisture may be excluded almost completely—an advantage for drugs sensitive toward oxidation and hydrolysis.

C. Molecular Solid Dispersions with Melt Extrusion Technology

Recently, the melt extrusion approach was introduced to prepare molecular solid dispersion systems. The breakthrough was achieved by the availability of a great variety of carrier systems of pharmaceutically approved excipients (51). Because of the surface activity of carriers used, a complete dissolution of drug from such solid dispersions can be obtained without the need for further pulverization, sieving, and mixing with excipients.

The essential advantages of the melt extrusion process are its solvent-free formation process and the dust-free processing environment (52). Because with solvent processes there are various problems relating to their use (environmental pollution, explosion proofing, and residual organic solvent), measures to counteract these problems are desirable. The same holds for contamination with dust. In addition, the melt extrusion process offers the chance to convert drugs to the amorphous state or more often to dissolve the drugs in the matrix polymer. Therefore it is capable of handling actives of different particle sizes, as well as amorphous solids or other polymorphic forms leading to the same product.

A number of solid dispersions have been developed using the melt extrusion process with a drug load ranging from 30% to 60% with real time stability up to 7 years (Table 1). During this time period, no crystallization could be detected by means of XRPD or differential scanning calorimetry (53). It seems clear that the stability of molecular solid dispersions is achieved when the solubility of a given drug in the carrier is not exceeded. By definition, such systems are thermodynamically stable as their stability is related to the solubility of the active and not to the viscosity of the matrix as in super-saturated systems, which are kinetically stable.

Therefore to secure storage stability at elevated temperatures of kineti-cally stabilized systems, compositions with glass transition temperatures significantly above the storage conditions would generally be preferable. To be cautious, however, a whole range of analytical methods to determine the absence or the presence of crystalline drug substance in the pharmaceutical product is currently applied to further prove the quality and the stability of the solid dispersion products (Table 2).

Table 1 Stability Data of Solid Dispersions Manufactured by the Melt Extrusion Process

Drug substance	Melting point [°C]	Approximate stable period proved to date [years]
Furosemide	206	5
Nifedipine	172–174	7
Paracetamol	169	4

Table 2 Analytical Methods Applied to Examine Solid Dispersions

---

- Hot stage microscopy
- X-ray diffraction [55]
- Thermal analysis/Differential scanning calorimetry [56]
- Near infrared spectroscopy [57]
- Microcalorimetry [58]

In addition to oral dosage forms, bioadhesive ophthalmic inserts (59), implants, and topical films (60) have been developed.

Repka et al. (61) also showed that a thermally labile drug, hydrocortisone, could be successfully incorporated into HPC films produced by hot melt extrusion. Aitken-Nichol et al. (62) investigated the viability of hot melt extrusion technology in 1996 for the production of thin, flexible, acrylic films for topical drug delivery. These researchers showed that lidocaine HCl was able to plasticize the acrylic polymer and that the drug was completely dispersed at the molecular level in the extruded films. Solubilized drug molecules were shown to plasticize the polymer by increasing the average polymer chain spacing.

In Vivo Data

Along with literature describing the in vitro performance of solid dispersions, a number of melt-extruded molecular dispersions have been examined in humans.

The antiretroviral loviride has been melt-extruded to a solid molecular dispersion in HPMC and has shown remarkably lower food effect compared to capsules. Area under the curve (AUC) values are given from a study of healthy volunteers (63).

Antifungal compositions of itraconazole were prepared as molecular solid dispersions from the melt extrusion process. In a limited number of volunteers, the melt-extruded tablet gave an AUC value of itraconazole in the fasted state that was 2.3 times the AUC of the marketed reference capsule (64).

Interestingly, melt-extruded dispersions of ibuprofen gave equivalent data to the lysinate salt of ibuprofen in healthy volunteers. More precisely, bioequivalence was demonstrated with the relevant parameters AUC and

C_{max}. Also, the t_{max} as a measure for onset proved to be equivalent to 0.5 hr for test and reference (65).

D. Regulatory Aspects

Products from the melt extrusion process have been approved in the United States and European countries. The process technology lends itself to comprehensive documentation, thus satisfying regulatory authorities. It is a major advantage that extrusion is a mature engineering technology. As a process, it provides many parameters, such as feeding rate, segmental temperatures, and pressure or applied vacuum, which can be monitored on-line with local meters and sensors. Such data contribute to the comprehensive documentation and the quality of production lots and may finally simplify quality control.

IV. CONCLUSION

Melt extrusion technology has opened a pathway to the efficient manufacture of drug delivery systems. Along with the use of dispersed, amorphous, and molecularly dissolved systems, a number of other formulation techniques can be applied using the melt extrusion approach: in situ salt formation, particle size homogenization, and tailored release profiles reveal the potential of melt extrusion as a drug delivery technology. Unlike other delivery systems, most of which consist of a random dispersion of a drug in a matrix, extrusion technology may be used in the future to place exact amounts of active in well-defined layers.

REFERENCES

1. Atkinson RM, Bedford C, Child KJ, Tomich EG. Effect of particle size on blood griseofulvin levels in man. Nature 1962; 193:588–589. Levy J. Lancet 1962; 2:723–724.
2. Sekiguchi K, Obi N. Studies on absorption of eutectic mixtures. 1. A comparison of the behaviour of eutectic mixtures of sulfathiazole and that of ordinary sulfathiazole in man. Chem Pharm Bull 1961; 9:866–872.
3. Hajratwala BR. Dissolution of solid dispersion systems. Aust J Pharm Sci 1974; 4:101–109.

4. Bloch DW, Speiser PP. Solid dispersions—fundamentals and examples. Pharm Acta Helv 1987; 62:23–27.

5. Hempenstall J. Latest developments in tablet and capsule formulation. London: European Continuing Education College, Master Classes in Solid Dosage Forms, 1997:1–14.

6. Serajuddin AT. Solid dispersion of poorly water-soluble drugs: early promises, subsequent problems, and recent breakthroughs. J Pharm Sci 1999; 88:1058–1066. Leuner C, Dressman J. Improving drug solubility for oral delivery using solid dispersions. Eur J Pharm Biopharm 2000; 50:47–60.

7. Chiou WL, Riegelman S. Pharmaceutical applications of solid dispersion systems. J Pharm Sci 1971; 60:1281–1301.

8. Breitenbach J. Two Concepts, One Technology: Controlled Release and Solid Dispersion with Meltrex™. Modified Release Drug Delivery Technology. New York: Marcel Dekker, 2001:125–134.

9. Doherty C, York PJ. Evidence for solid- and liquid-state interactions in a furosemide–polyvinylpyrrolidone solid dispersion. Pharm Sci 1987; 76: 731–737.

10. Fukuoka E. Glassy state of pharmaceuticals: III. Thermal properties and stability of glassy pharmaceuticals and their binary glass systems. Chem Pharm Bull 1989; 37:1047–1050.

11. Etter MC. Hydrogen bond directed CO-crystallization and molecular recognition properties of diarylureas. J Am Chem Soc 1990; 112:8415–8426.

12. Hancok BC, Shamblin SL, Zografi G. Molecular mobility of amorphous pharmaceutical solids below their glass transition temperatures. Pharm Res 1995; 12:799–806. De Filippis P. Dissolution rates of different drugs from solid dispersions with Eudragit RS. Eur J Pharm Sci 1995; 3:265–271.

13. Breitenbach J, Schrof W, Neumann J. Confocal Raman-spectroscopy: analytical approach to solid dispersions and mapping of drugs. Pharm Res 1999; 16: 1109–1113.

14. Craig DQM. Polyethylene glycols and drug release. Drug Dev Ind Pharm 1990; 16:2501–2526. Corrigan OI. Mechanisms of dissolution of fast release solid dispersions. Drug Dev Ind Pharm 1985; 11:697–724.

15. Corrigan OI. Mechanism of dissolution of fast release solid dispersions. Drug Dev Ind Pharm 1985; 11:697–724.

16. Tantishaiyakul V, Kaewnopparat N, Ingkatawornwong S. Properties of solid dispersions of piroxicam in polyvinylpyrrolidone. Int J Pharm 1999; 181:143–151.

17. Zingone G, Moneghini M, Rupena P, Vojnovic D. Characterization and dissolution study of solid dispersions of theophylline and indomethacin with PVP/VA copolymers. STP Pharma Sci 1992; 2:186–192.

18. Follonier N, Doelker E, Cole ET. Various ways of modulating the release of diltiazem hydrochloride from hot-melt extruded sustained release pellets prepared using polymeric materials. J Control Release 1995; 36:243–250.

19. Yano K, Kajiyama A, Hamada M, Yamamoto K. Constitution of colloidal particles formed from a solid dispersion system. Chem Pharm Bull 1997; 45:1339–1344.
20. Abd A, El-Bary, Geneidi AS, Amin SY, El-Ainan AA. Preparation and pharmacokinetic evaluation of carbamazepine controlled release solid dispersion granules. J Drug Res Egypt 1998; 22:15–31.
21. Nakagami H. Solid dispersions of indomethacin and griseofulvin in non-porous fumed silicon dioxide, prepared by melting. Chem Pharm Bull 1991; 39:2417–2421.
22. Ford JL. The current status of solid dispersions. Pharm Acta Helv 1986; 61:69–88.
23. Lefebvre C, Brazier M, Robert H, Guyot-Hermann AM. Solid dispersions, why and how? Industrial aspect. STP Pharma 1985; 4:300–322.
24. Doelker M, El-Egakey MA, Soliva M, Speiser P. Hot extruded dosage forms. Pharm Acta Helv 1971; 46:31–52.
25. Vervaet C, Baert L, Remon JP. Extrusion–spheronisation—a literature review. Int J Pharm 1995; 116:131–146.
26. Reference deleted.
27. Follonier N, Doelker E, Cole ET. Various ways of modulating the release of diltiazem hydrochloride from hot-melt extruded sustained release pellets prepared using polymeric materials. J Control Release 1995; 36:243–250.
28. Follonier N, Doelker E, Cole ET. Evaluation of hot-melt-extrusion as new technique for the production of polymer based pellets for sustained release capsules containing high loadings of freely soluble drugs. Drug Dev Ind Pharm 1994; 20:1323–1339.
29. Wacker S, Soliva M, Speiser P. Injection moulding as a suitable process for manufacturing solid drug dispersion or solutions. J Pharm Sci 1971; 60:1281–1300.
30. European Patent 0406315, 1988.
31. Zhang F, McGinity JW. Properties of sustained-release tablets prepared by hot-melt extrusion. Pharm Dev Technol 1999; 4:241–250.
32. U.S. Patent 6071539, 1997.
33. Sarraf AG, Barja F, Quali L, Sarraf EG, Gurny R, Doelker E. Effect of polymer orientation induced by melt extrusion on drug release. Proceedings of International Symposium on Controlled Release Bioactive Materials, San Diego, 2001.
34. Nakamichi K, Yasuura H, Fukui H, Oka M, Izumi S. Evaluation of a floating dosage form of nicardipine hydrochloride and hydroxypropylmethylcellulose acetate succinate prepared using a twin-screw extruder. Int J Pharm 2001; 218:103–112.
35. Hülsmann S, Backensfeld T, Keitel S, Bodmeier R. Melt extrusion—an alternative method for enhancing the dissolution rate of 17-estradiol hemi-hydrate. Eur J Pharm Biopharm 2000; 49:237–242.

36. European Patent 1011640, 1997.
37. European Patent 0998304, 1997.
38. European Patent 0686392, 1995.
39. Repka MA, Gerding TG, Repka SL, McGinity JW. Influence of plasticizers and drugs on the physical–mechanical properties of hydroxypropylcellulose films prepared by hot melt extrusion. Drug Dev Ind Pharm 1999; 25:625–633.
40. Taylor LS, Zografi G. Sugar–polymer hydrogen bond interactions in lyophilized amorphous mixtures. J Pharm Sci 1998; 87:1615–1621.
41. European Patent 0003612, 1999.
42. Matsumoto T, Zografi G. Physical properties of solid molecular dispersions of indomethacin with poly(vinylpyrrolidone) and poly(vinylpyrrolidone-co-vinyl-acetate) in relation to indomethacin crystallization. Pharm Res 1999; 16:1722–1728.
43. Forster A, Hempenstall J, Rades T. Characterization of glass solutions of poorly water-soluble drugs produced by melt extrusion with hydrophilic amorphous polymers. J Pharm Pharmacol 2001; 53:303–315.
44. Yuasa H, Ozeki T, Kanaya Y, Oishi K. Application of the solid dispersion method to the controlled release of medicine: IV. Precise control of the release rate of a water soluble medicine by using the solid dispersion method applying the difference in the molecular weight of a polymer. Chem Pharm Bull 1993; 41:933–936. Tantishaiyakul V, Kaewnopparat N, Ingkataworrnwong S. Properties of solid dispersions of piroxicam in polyvinylpyrrolidone. Int J Pharm 1999; 181:143–151.
45. Hamaura T, Newton JM. Interaction between water and poly(vinylpyrrolidone) containing polyethylene glycol. J Pharm Sci 1999; 88:1228–1233.
46. Ozeki T, Yuasa H, Kanaya Y. Application of the solid dispersion method to the controlled release medicine: IX. Difference in the release of flurbiprofen from solid dispersions with poly(ethylene oxide) and hydroxypropylcellulose and the interaction between medicine and polymers. Int J Pharm 1997; 155:209–217.
47. Khan MA, Shojaei AH, Karnachi AA, Reddy IK. Comparative evaluation of controlled-release solid oral dosage forms prepared with solid dispersions and coprecipitates. Pharm Technol 1999; 5:58–72.
48. Thakkar AL, Hirsch CA, Page JG. Solid dispersion approach for overcoming bioavailability problems due to polymorphism of nabilone, a cannabinoid derivative. J Pharm Pharmacol 1977; 29:784.
49. European Patent 0580860, 1991.
50. Nakamichi K, Yasuura H, Kukui H, Oka M, Izumi S, Andou T, Shimizu N, Ushimaru K. New preparation method of solid dispersion by twin screw extruder. Pharm Technol Jpn 1996; 12:715–729.
51. Breitenbach J, Berndl G, Neumann J, Rosenberg J, Simon D, Zeidler J. Solid dispersions by an integrated melt extrusion system. Proceedings of International Symposium on Controlled Release Bioactive Materials, Las Vegas, 1998:804–805. Breitenbach J. Feste Lösungen durch Schmelzextrusion—ein integriertes Herstellkonzept. Pharm Unserer Zeit 2000; 29:1–5.

52. Berndl G, Breitenbach J, Neumann J, Reinhold U, Rosenberg J, Vollgraf C, Zeidler J. Polymer/drug-melt extrusion: control of drug release and morphology in pharmaceutical formulations. American Association of Pharmaceutical Scientists, Annual Meeting Abstracts, San Francisco, 1998:296. Nakamichi K, Yasuura H, Fukui H, Oka M, Izumi S, Andou T, Shimizu N, Ushimaru K. A process for the manufacture of nifedipine hydroxypropylmethylcellulose phthalate solid dispersions by means of a twin screw extruder and appraisal thereof. Yakuzaigaku 1996; 56:15–22.

53. Grünhagen HH. Polymer/drug-melt extrusion: therapeutic and technological appeal. Pharm Technol Eur 1996; 8:22–28.

54. Taylor LS, Langkilde FW, Zografi G. Fourier transform Raman spectroscopic study of the interaction of water vapor with amorphous polymers. J Pharm Sci 2001; 90:888–901.

55. de Villiers MM, Wurster DE, Van der Watt JG, Ketkar A. X-ray powder diffraction determination of the relative amount of crystalline acetaminophen in solid dispersions with polyvinylpyrrolidone. Int J Pharm 1998; 163:219–224.

56. Giron D. Contribution of thermal methods and related techniques to the rational development of pharmaceuticals, Part 1. PSTT 1998; 1:191–199. Giron D. Contribution of thermal methods and related techniques to the rational development of pharmaceuticals, Part 2. PSTT 1998; 1:262–268.

57. Brittain HG. Spectral methods for the characterization of polymers and solvates. J Pharm Sci 1997; 86:405–412.

58. Phipps MA, Mackin LA. Application of isothermal microcalorimetry in solid state drug development. PSTT 2000; 3:9–17.

59. Baeyens V, Kalsatos V, Boisrame B, Fathi M, Gurny R. Evaluation of soluble Bioadhesive Ophthalmic Drug Inserts (BODI) for prolonged release of gentamicin: lachrymal pharmacokinetics and ocular tolerance. J Ocul Pharmacol Ther 1998; 14:263–272.

60. Repka MA, McGinity JW. Bioadhesive properties of hydroxypropylcellulose topical films produced by hot-melt extrusion. J Control Release 2001; 70(51): 341–351.

61. Repka MA, Gerding TG, Repka SL, McGinity JW. Influence of plasticizers and drugs on the physical–mechanical properties of hydroxypropylcellulose films prepared by hot-melt extrusion. Drug Dev Ind Pharm 1999; 25:625–633.

62. Aitken-Nichol C, Zhan F, McGinity JW. Hot melt extrusion of acrylic films. Pharm Res 1996; 13:804–808.

63. European Patent 0872233, 1997.

64. European Patent 0904060, 1997.

65. European Patent 1001797, 1997.

14

Melt-Extruded Particulate Dispersions

S. Craig Dyar and Matthew Mollan
Pfizer, Inc., Ann Arbor, Michigan, USA

Isaac Ghebre-Sellassie
*MEGA Pharmaceuticals, Asmara, Eritrea, and Pharmaceutical Technology
Solutions, Morris Plains, New Jersey, USA*

I. INTRODUCTION

One of the most significant problems facing the pharmaceutical scientist relates to enhancing the solubility of the increasing number of poorly water-soluble compounds being brought into the development pipeline. This large number is considered the result of high throughput screening methods, which tend to focus on the target mechanism at the expense of other desired properties, such as solubility. In general, poorly water-soluble drugs that undergo dissolution rate-limited gastrointestinal absorption show an increased bioavailability when the rate of dissolution is improved (1). A number of techniques have been utilized to enhance the dissolution rate and potentially the bioavailability of poorly water-soluble drugs. These techniques include relatively simple processes such as particle size reduction (2), salt selection (3), and the use of a cosolvent (3) or surface-active agent (3). Endeavors that are more scientifically complex have also been attempted with varying degrees of success like the formation of a molecular complex (4) or solid dispersion (5). The use of a metastable polymorphic form has also been evaluated (6). In solid dosage form development, the most commonly used technique has been particle size reduction because it is one of the simplest techniques available to enhance the dissolution rate, and hence bioavailability, without bringing about significant polymorphic changes. Because the manufacturing of melt-extruded particulate dispersions, which is the subject of this chapter, utilizes particle size reduction as a first step, it is essential to discuss in detail the most commonly used size reduction techniques to appreciate the relationships.

A review of the literature indicates that there are several instances where particle size reduction leads to an improved dissolution rate and a bioavailability of poorly soluble compounds. One widely cited example is the micronization of griseofulvin that led to a dramatical increase in blood levels when compared with nonmicronized griseofulvin. Other examples where micronization led to improved bioavailability include aspirin, tetracycline, and sulfadiazine. Naturally, as with any technique, there are limits to how much, if any, bioavailability will be increased. This will be discussed in a later section.

II. PARTICLE SIZE REDUCTION PROCESSES

A. Air Jet Milling

The air jet mill, also called a fluid energy mill or air-classifying mill, comprises a feed throat connected to a circular chamber with a number of air jets on the outer circumference of the chamber. These jets are aligned to provide air movement in a circular motion that spirals downward toward the center of the chamber. Particle reduction occurs due to collisions between the drug particles. The smaller particles move to the center of the chamber and are collected through an orifice in the bottom center. The feed rate, in conjunction with the nozzle pressure, determines the particle size of the milled drug. Particle sizes in the 1–30 μM range are possible with the air jet mill. There are several important advantages to using air jet mills, such as ease of cleaning, minimal sample contamination issues, and simplicity of sample collection.

B. Ball Milling

A ball mill consists of a cylindrical vessel and grinding media composed typically of spherical ceramic or steel balls. The drug particles are reduced to the nanometer size via attrition, compression, and impaction. The size, shape, density, and hardness of the grinding media also have an effect on particle size. The ratio of grinding media to powder, along with the fill level in the vessel, influences the degree of milling. The major disadvantages of ball milling processes involve requirements for the separation of the grinding media from the milled material and the potential contamination of the material from the grinding media.

C. Hammer Milling

The *Hammer mill* is widely used in the pharmaceutical industry. Powder is fed through a feed hopper into the milling chamber where hammers impact the material causing the particles to fracture. The particles rotate in the chamber until they are sufficiently small enough to pass through the screen at an angle. Hammer speed and shape, screen size, and feed rate affect the final particle size of the drug. The final particle size can be in the range of 20–50 μM. A variation of this mill uses a knife or cutting blade in place of a hammer to allow the milling of fibrous or elastic material.

Liquid nitrogen is sometimes added to the size reduction processes discussed above to overcome the limitations of standard milling operations, particularly when materials that have low transition temperatures are processed. The liquid nitrogen lowers the temperatures of the materials and increases their brittleness, thereby facilitating the milling process.

D. Supercritical Fluid

Supercritical fluid-based processing techniques have been used to reduce particle size in various nonpharmaceutical industrial processes, such as the paint industry for a number of years. The pharmaceutical industry has recently begun to investigate supercritical fluid technology in particle size reduction processes (7). The technique is capable of producing micron-size particles with a narrow size distribution and a low aspect ratio. It is also environment-friendly if carbon dioxide is employed as a medium. A major drawback of this technology is its higher cost relative to other size reduction techniques.

E. Homogenization

In high-pressure *homogenization*, two streams of drug particles suspended in liquid are directed at each other under high pressure to bring about high-velocity collisions among drug particles, resulting in particles that are in the micron level or below. As expected, the product is a suspension rather than fine powder. However, the suspension could be sprayed onto formulation components and dried to generate granulations that are suitable for solid dosage form development.

F.   Limitations of Size Reduction Processes

Size reduction processes and the handling of very fine powders have a number of issues associated with them. As a result, many researchers have attempted to resolve these issues, and their approaches will be discussed.

One potential problem is particle agglomeration, which occurs due to the static charge commonly generated during particle size reduction processes and associated cohesive forces that bind fine particles together (8). Agglomeration does not only impede powder flow during processing, but it could also potentially result in a decrease in dissolution rates and hence the bioavailability of drug products, provided that absorption is not the rate-limiting step. The technology developed by NanoSystems, now part of Elan, attempts to alleviate this problem by stabilizing the nanocrystals in a suspension using a bioadhesive or temperature-responsive polymer (9–11). Danazol is a poorly water-soluble compound (10 μg/ml) that was used to illustrate the applicability of the NanoSystems approach. The crystalline drug was milled in an aqueous medium to less than 200 nm and stabilized in a suspension to prevent agglomeration in the gastrointestinal tract. A study that compares the oral bioavailability of danazol from a NanoSystems suspension and an aqueous suspension of danazol particles was performed in dogs. The bioavailability of the drug from the NanoSystems danazol suspension was significantly higher than that from the aqueous danazol suspension, i.e., 82.3% vs. 5.1% (12). One concern with the suspension approach is Oswald ripening, which is a result of crystal growth caused by the dissolution of smaller particles followed by the precipitation of the drug molecules on the surface of larger particles (13). Substances, such as polyvinylpyrrolidone (PVP), casein, and polyvinylalcohol, have been successfully used to inhibit crystal growth.

Another issue is that the high amount of energy used in these milling processes can lead to significant temperature increases, thereby exposing the drug to very high temperatures. Degradation can occur if the decomposition temperature of the drug is exceeded. It is also likely that the drug could be converted to an amorphous form or another crystalline polymorph during milling (14,15). Both amorphous and polymorphic conversions can result in decreased stability.

G.   Surfactants and Wetting Agents

The addition of a wetting agent, such as a surfactant or water-soluble polymer, has been shown to improve the dissolution rate in some drug formulations. A common surfactant used in pharmaceutical preparations is polysorbate-80

(Tween-80). When coupled with size reduction, wetting agents could have an additive or synergistic effect that would further improve the dissolution rate of poorly soluble compounds.

III. SOLID DISPERSIONS

Solid dispersions have been defined in the literature as dispersions of one or more active ingredients in inert carriers or matrices in the solid state, prepared by the fusion, solvent, or fusion–solvent methods (16). The same authors have also described several physical forms of solid dispersions, such as simple eutectics, solid solutions, glass solutions, glass suspensions, and amorphous precipitations, in a crystalline carrier or combinations thereof. Other authors have defined solid dispersions as intimate mixtures of drug substances (solutes) and diluents or carriers (solvent or continuous phase) (17). Solid dispersions are classified based on drug solubility in the carrier. Those cases where drug substances are dispersed at the molecular level are known as molecular dispersions or solid solutions, while those dispersed at the particulate level are known as particulate dispersions or solid suspensions. Melt-extruded molecular dispersions or solid solutions are discussed extensively in Chapter 14 and will be addressed briefly in this chapter. An extremely important factor to consider when developing a solid solution is the solid-state solubility of the drug in the solid solution carrier. Ideally, the drug will have unlimited solubility in the carrier, and any drug/carrier ratio results in a solid solution. The worst-case scenario occurs when the drug has limited solubility in the carrier, leading to a low drug/carrier ratio, which results in a large dosage unit. The stability of the formulation is extremely dependent on the interaction between the drug and the carrier. The weaker the interaction and the closer the concentration of the solution is to saturation, the higher the likelihood of chemical or physical instability of the formulation.

IV. PARTICULATE DISPERSIONS

Particulate dispersions have been manufactured in at least three different ways. Two of these processes will be addressed briefly, while the third process, which keeps the crystallinity of drug in its original form during processing, is the main topic of this chapter and will be discussed extensively. One method involves a two-step process that converts, as a first step, the crystalline drug to a solid solution and then forces the drug to revert to the crystalline form. The control of this conversion process, however, is not only difficult, but it also cannot assure a regeneration of the original polymorph, which could have stability implications. The other method is based on the solvent spray drying technique and generates particulate dispersions via the solid solution route. Two major problems associated with this method are the need to use solvents and that the resultant spray-dried material generally has low bulk density. The use of solvents is tightly regulated in many U.S. locations, rendering the process extremely expensive. The lower density of the spray-dried dispersion often results in problems with blending and tableting, and may necessitate the need to roller compact the material to improve the flow and the content uniformity of the final dosage forms.

A. Melt-Extruded Particulate Dispersions

In this process, the drug and the excipients are blended and introduced into a melt extruder or mixing bowl. The components are then mixed in the molten state at appropriate temperatures, in which the excipients become soft or even melt but the drug remains solid. The drug particles are embedded in the excipient matrices. The excipient(s) should have a melting or softening point below the melting point of the drug. The process can be used to manufacture controlled-release particulate dispersions (18) as demonstrated by Badische Anilin- & Soda-Fabrik AG (BASF), or to enhance the solubility of poorly soluble compounds as will be discussed in the following sections.

The improvement of the dissolution rates of poorly water-soluble compounds is achieved by embedding fine crystals in a water-soluble polymeric material under appropriate temperatures and pressures followed by the size reduction of the congealed melt. Basically, very fine drug particles are dispersed within polymeric carriers to form particulate dispersions. Water-soluble polymers such as polyvinylpyrrolidone, hydroxypropyl cellulose (HPC), or hydroxypropyl methylcellulose (HPMC) are commonly used carriers. In an ideal system, all the drug particles are "coated" by the water-soluble polymers to produce products that have improved dissolution rates (19). Just as in molecular dispersions, the polymer/drug ratio is very important. The drug solubility in the polymer directly influences the type of system that can be developed. In contrast to the situation with solid solutions, the drug should have minimal solubility in the polymer. The amount of polymer necessary to "coat" the particles in melt-extruded particulate dispersions is much lower than that required for molecular dispersions due to surface area considerations and differences in the functionalities of the polymers in the two technologies. In melt-extruded molecular dispersions, the role of the polymer is to act as a solubilizer through molecular associations. While in melt-extruded particulate dispersions, the polymer intersperses itself between adjacent drug crystals to improve wettability. Unlike molecular dispersions where the drug exists in a thermodynamically unstable amorphous form, particulate dispersions have the drug in the more thermodynamically stable crystalline form. The improved stability of a particulate dispersion over a solid solution may come at the cost of a slower dissolution rate, and possibly lower bioavailability. However, there are many processes and systems in the body that may affect bioavailability, and the only way to determine the impact is to perform a clinical trial.

As stated earlier, particulate dispersions are more stable than solid solutions and consequently have less stringent packaging and storage requirements, which in turn could lead to appreciable cost savings. In addition, particulate dispersions do not require drugs to be processed at temperatures at or above their melting points. This offers significant benefit over solid solutions if the drug candidate has the propensity to decompose during elevated processing temperatures.

B. Practical Applications

In this section, an attempt will be made to describe the applicability of particulate dispersions in the formulation of poorly water-soluble drug substances using specific examples. Detailed process descriptions and physicochemical characterizations of the dispersions will be presented.

Example A

In this example, an investigational drug, INV-A, that is practically water-insoluble in the gastrointestinal pH range of 1.0–7.5 is discussed. The drug has an aqueous solubility of <1 μg/ml with a corresponding bioavailability of 15% in dogs. Prior to this study, INV-A was formulated as a solid solution by melt extrusion methods in order to enhance its dissolution rate and oral bioavailability. The dissolution rate was increased 28-fold over the bulk drug alone and resulted in a 2.5-fold increase in dog bioavailability. The results indicated that the bioavailability of the drug is dissolution rate-limited and can be enhanced by formulation techniques. As a result, INV-A was selected as a model drug in determining whether the dissolution rate of poorly water-soluble drugs could be improved by the formulation of particulate dispersions.

Process Description. A Brabender mixing bowl was used to prepare the INV-A particulate dispersions. The mixing bowl is a heated high-shear mixer that has been extensively used in the plastic industry to identify processing temperatures for polymer formulations (Figure 1). The sample requirement is only 30–60 g of material; hence, a number of experiments can be performed in a short period with a relatively small amount of material. Initially, a series of experiments were conducted to determine the optimum processing temperature ranges for the polymers because it is the transition temperature of the polymers that dictates the final processing conditions. Then the milled drug was blended with the appropriate type and quantity of polymers, and introduced into the mixing bowl. The formulation components were then mixed in the molten state under predetermined processing conditions of shear and temperature. The formulation composition and processing temperatures are given in Table 1. The conditions were sufficient to melt or to soften the polymers, but not to melt the active ingredient. The resulting products were collected, milled, and sieved.

Characterization.   A sieve cut of 80–100 mesh was used for the dissolution studies and other tests. The crystallinity of the INV-A in the particulate dispersions was characterized by powder x-ray diffraction and polarizing optical microscopy. The peaks that are characteristic of bulk INV-A were present at the same location as the samples processed at 110°C (Figure 2).

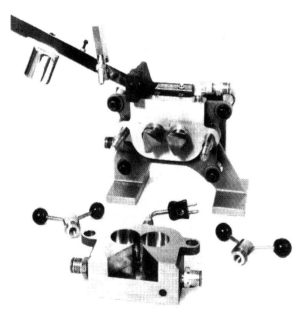

Figure 1   Mixing bowl with mixers.

Table 1 Particulate Dispersion Composition and Processing Temperature

INV-A (%)	PEG (%)	HPC (%)	PVP (%)	Processing temperature (°C)
80	10	–	10	110
80	10	10	–	110
75	10	–	15	130
75	10	15	–	130
75	5	20	–	130
75	0	–	–	130

Dissolution studies were performed at pH 8 with sodium lauryl sulfate (SLS). The results shown in Figure 3 clearly indicate that all the INV-A particulate dispersions exhibited a greater rate and extent of dissolution than those of the bulk INV-A and the physical mixture of the drug and the polymer. The INV-A/HPC (75/25) particulate dispersion exhibited the fastest dissolution rate of the samples studied. This is likely due in part to the higher polymer/drug ratio. This sample also had the highest initial dissolution rate, which was about 12 times higher than bulk INV-A.

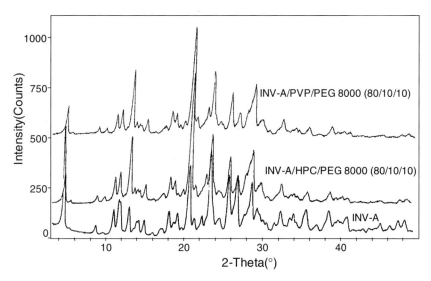

Figure 2 X-ray powder diffractograms of INV-A and its PEG 8000 containing particulate dispersions.

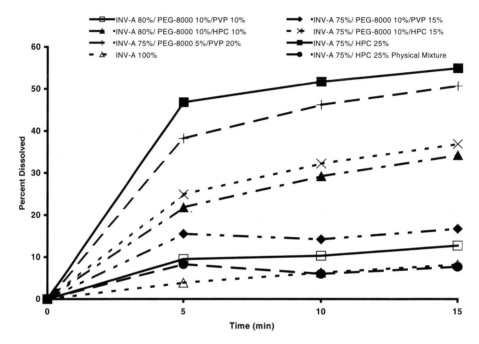

Figure 3 Comparative dissolution profiles of various INV-A/PEG-8000/HPC particulate dispersions, Pure INV-A, and INV-A/HPC (75/25). Physical mixture in pH 8 (0.05 M) phosphate buffer (900 ml of medium at 37°C, 75 rpm paddle, UV method of drug analysis at 284 nm).

Example B

In this example, another poorly water-soluble investigational compound, INV-B, which has an aqueous solubility of <1 µg/ml and a melting point of 179.4°C, is described. The drug was micronized to produce a material that had a particle size fraction of 90% less than 4.16 µm.

Process Description. The micronized drug was first blended and processed in a DACA Micro-Compounder to generate the particulate dispersions. The formulations are shown in Table 2. The DACA uses conical corotating twin screws to mix and to convey the material under controlled temperatures and pressures (Figure 4). The system was heated to 150°C, a temperature that allowed the polymer to melt or to soften while the drug remained in the crystalline state. The extrudates were cooled at room temperature and milled using the SPEX 6800 freezer mill to provide a particle size distribution of less

Table 2 Formulation Composition

	Amount (%)			
Description	INV-B	PVP	HPMC-AS L	Tween 80
INV-B/HPMC-AS	60	–	40	–
INV-B/HPMC-AS/Tween 80	60	–	39	1
INV-B/PVP	60	40	–	–
INV-B/PVP/Tween 80	60	39	–	1

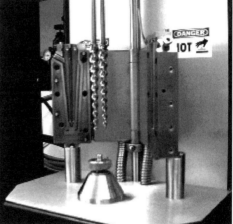

Figure 4 The DACA Mini-Compounder with close-up view of barrel and screw section.

than 44 μm. This mill uses magnetically driven impactors contained in polycarbonate cylinders to reduce the particle size of the material. The cylinders are submerged in a liquid nitrogen bath. All formulations processed well in the microcompounder. Finally, size 2 capsules were filled with a blend of 250 mg of milled material (equivalent to 150 mg of active material) and 7.5 mg of the disintegrant, croscarmellose sodium, for dissolution testing and biostudies.

Characterization. The hydroxylpropyl methylcellulose acetate succinate (HPMC-AS) formulations reached a maximum percent dissolved at 45 min and were about two to seven times higher than those of formulations containing PVP. The HPMC-AS formulations also provided about three to four times higher percent dissolved than those of the micronized drug substance at 45 min (Figure 5). The powder x-ray diffractograms showed

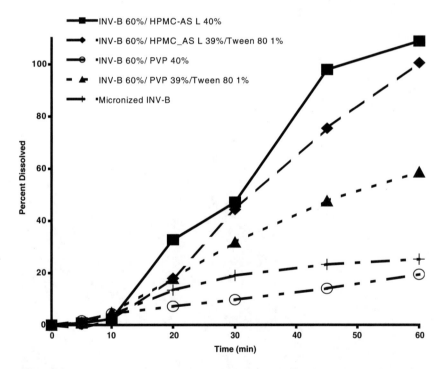

Figure 5   The percent dissolved in pH 6.8 phosphate buffer dissolution medium with 0.5% Tween 80.

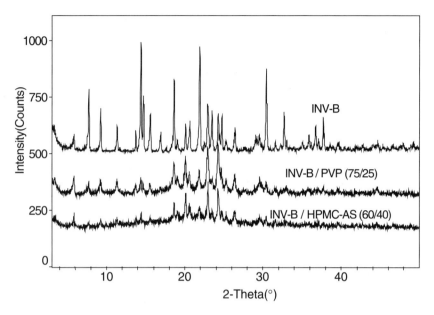

Figure 6 X-ray powder diffractograms of INV-B and its non-Tween 80 containing particulate dispersions.

Table 3 Dog Pharmacokinetic Results ($n=3$)

Description	Dose (mg/kg)	C_{max} (µg/ml)	T_{max} (hr)	AUC $(0-\infty)$ (µg[a] hr/ml)	F (%)
Capsule with PVP solid dispersion	15	0.069	1.7	0.165	1.0
Capsule with 40% HPMC-AS L particulate dispersion[a]	15	0.268	1.0	1.11	6.7
Capsule with 1% Tween 80/39% HPMC-AS L particulate dispersion[a]	15	0.261	2.0	0.984	6.0

[a] Exposure in one dog was very low and the area under the curve (AUC) could not be calculated. Thus, the bioavailability reported was derived from two dogs.

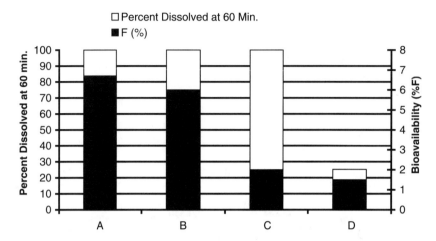

Figure 7   The percent dissolved in pH 6.8 phosphate buffer dissolution medium with 0.5% Tween 80 vs. the percent bioavailable (F) in dogs of formulation. (A) Capsule containing 60% micronized INV-B and 40% HPMC-AS L as a particulate dispersion; (B) capsule containing 60% micronized INV-B, 1% Tween 80, and 39% HPMC-AS L as a particulate dispersion; (C) capsule containing 60% INV-B and 40% PVP as a solid dispersion; and (D) capsule containing 100% micronized INV-B.

that the formulations retained significant crystallinity after processing (Figure 6) when compared to the bulk drug substance. The pharmacokinetic results from the dog study are shown in Table 3 and Figure 7. Although this was a relatively small study, the results indicate that the particulate dispersion preparations resulted in up to a sevenfold increase in bioavailability (F). The C_{max} was two to four times higher and the T_{max} was equivalent to, or greater than, those of the other formulations.

V. SUMMARY

Enhancing the dissolution rates, and hence the bioavailability of poorly water-soluble drug substances, is an area that has been thoroughly investigated. While conventional size reduction processes have been applied successfully in certain instances to overcome the problem, they have not been able to satisfy the need. Some of these shortcomings relate to processing issues while others could not achieve the desired enhancement in dosage form performance. As a result, additional innovative formulation and processing approaches that allow the routine formulation of poorly soluble active ingredients have been explored. Melt-extruded particulate dispersions, which have been discussed extensively in this chapter, are expected to overcome most of the drawbacks of existing technologies. Cohesive powders, which tend to agglomerate and to impede dissolution rates, can be formulated to generate dosage forms that have acceptable bioavailability. The physical and chemical instabilities observed with molecular dispersions or solid solutions are absent in particulate dispersion formulations because the formulated drug exists in its thermodynamically stable crystalline state. While other processes could be used to manufacture particulate dispersions, the melt extrusion process is the preferred method due to its inherent intense mixing capabilities and well-controlled processing parameters. It is also continuous and presents minimal scale-up issues. In addition, the process does not involve the use of solvents, which makes it very cost-effective and environment-friendly.

Another important aspect that is critical to the formation of particulate dispersions is a thorough characterization of the materials. Unlike bulk drug substances, particulate dispersions are complex matrices from which it is difficult to obtain conclusive results with one analytical technique. Therefore when characterizing particulate dispersions, it is essential to utilize a combination of techniques to establish the validity of the findings. One of these methods is dissolution testing—the ultimate predictor of in vivo performance. While other techniques may provide the physical state of the dispersions, dissolution tests provide a much better indication of how the particulate dispersion may perform in a test subject, especially when working with physically relevant dissolution media and conditions.

REFERENCES

1. Mayersohn M. Principles of drug absorption. In: Banker GS, Rhodes C, eds. Modern Pharmaceutics. Drug and the Pharmaceutical Sciences. Vol. 72. 3rd Ed. New York: Marcel Dekker, 1996:21–73.
2. Fincher JH. Particle size of drugs and its relationship to abstivity. J Pharm Sci 1968; 57:1825–1835.
3. Gould PL. Salt selection for basic drugs. Int J Pharm 1986; 33:201–217.
4. Bekers O, Uijtendaal EV, Beijnen JH, Bult A, Underberg WJM. Cyclodextrins in the pharmaceutical field. Drug Dev Ind Pharm 1991; 17:1503–1549.
5. Chiou WL, Riegelman S. Pharmaceutical applications of solid dispersion systems. J Pharm Sci 1971; 62:1281–1302.
6. Haleblian JK, McCrone W. Pharmaceutical applications of polymorphism. J Pharm Sci 1969; 58:911–929.

7. Chattopadhyay P, Gupta RB. Production of griseofulvin nanoparticles using supercritical CO_2 antisolvent with enhanced mass transfer. Int J Pharm 2001; 228:19–31.

8. Carstensen J. Advanced Pharmaceutical Solids. New York, NY: Marcel Dekker, 2001:314–318.

9. Bagchi P, Scaringe RP, Bosch HW. Co-microprecipitation of nanoparticulate pharmaceutical agents with crystal growth modifiers. U.S. Patent 5,665,331.

10. Lan D. Method of preparing stable drug nanoparticles. U.S. Patent 5,534,270.

11. Wong S. Sugar base surfactant for nanocrystals. U.S. Patent 5,622,938.

12. Liversidge GG, Cundy KC. Particle size reduction for improvement of oral bioavailability of hydrophobic drugs: I. Absolute oral bioavailability of nano-crystalline danazol in beagle dogs. Int J Pharm 1995; 125(1):91–97.

13. Hem SL, Feldkamp JR, White JL. Basic chemical principles related to emulsion and suspension dosage forms. In: Lachman L, Lieberman HA, Kanig JL, eds. The Theory and Practice of Industrial Pharmacy, Philadelphia, PA: Lea and Febiger, 1986:116–117.

14. Otsuka M, Kaneniwa N. Effect of grinding on the transformations of polymorphs of chloramphenicol palmitate. Chem Pharm Bull 1985; 33(4):1660–1668.

15. Lee KC, Hersey JA. Crystal modification of methisazone by grinding. J Pharm Pharmacol 1977; 29:249–250.

16. Chiou W-L, Riegelman S. Pharmaceutical applications of solid dispersion systems. J Pharm Sci 1971; 60:1281–1302.

17. Nanda AR, Rowlings CE, Barker NP, Sheen P-C. Immediate release solid dispersions for oral drug delivery. In: Liu R, ed. Water-Insoluble Drug Formulation. Denver, CO: Interpharm Press, 2000:493–523.

18. Grabowski S, Sanner A, Rosenberg J. Compositions which contain active substances and are in the form of solid particles. U.S. Patent 5,641,516.

19. Ghebre-Sellassie I. Solid pharmaceutical dosage forms in form of a particulate dispersion. PCT 9908660.

15
Extrusion/Spheronization

David F. Erkoboni
FMC Corporation, Princeton, New Jersey, USA

I. INTRODUCTION

Extrusion/spheronization is a multiple-step process capable of making uniformly sized spherical particles commonly referred to as spheres or pellets. The pellets produced using this method can be used for both immediate-release or modified-release applications. They are typically filled into hard gelatin capsules, but can also be compressed into tablets. The major advantage of extrusion/spheronization over other methods of producing drug-loaded spheres or pellets is the ability to incorporate high levels of actives without producing an excessively large particle. Additionally, the process is more efficient than other techniques for producing pellets.

Spheronization is a process invented by Nakahara in 1964. The U.S. Patent (3,277,520) describes a "Method and Apparatus for Making Spherical Granules" from wet powder mixtures or granulations [1]. The equipment described in the patent was commercialized by Fuji Denki Kogyo Co. under the trade name Marumerizer®. It is generically referred to today as a spheronizer. The process went widely unnoticed in the pharmaceutical industry until 1970, when two articles were published by employees of Eli Lilly and Co. Conine and Hadley [2] described the steps involved in the process including (1) dry blending, (2) wet granulation, (3) extrusion, (4) spheronization, (5) drying, and (6) optional screening. Reynolds [3] went on to further describe the equipment and the mechanics of the process, including the movement of the particles within the spheronizer, which results in a characteristic ropelike formation. This phenomenon is an important indicator in the formulation and process development and will be discussed in greater detail. Both publications cite desirable attributes of spherical particles that can be achieved. These attributes include good flow, low dusting, uniform size distribution, low friability, high hardness, ease of coating, and reproducible packing. While it should be noted that the properties of spheres and pellets produced using this method can be drug-specific, they are generally reproducible even at high drug loading levels. Additionally, they offer the common therapeutic advantages seen with multiparticulate drug delivery systems, such as a reduced risk of dose dumping in modified-release products, as well as less gastrointestinal irritation [4]. From the publication of these articles up to today, the interest in extrusion/spheronization has continued to grow. The process has become an established pelletization technique with an increasing number of products currently being produced. In the interim, the industry interest was primarily driven by the academe. The increased popularity is, in part, due to the ease of processing and efficiency as compared to other common pelletization techniques such as spray or powder layering. There is a growing understanding of the effects of process parameters and material characteristics. However, as suggested by Newton [5], the preparation of pellets by extrusion/spheronization is still a technology without complete understanding.

II. APPLICATIONS

The use of multiparticulate spherical pellets as a system to deliver drugs offers a broad number of options and capabilities. The use of extrusion/spheronization to produce the pellets results in a robust product having unique properties compared to other pelletization techniques. Potential applications include both immediate release and controlled release. Two or more actives can easily be combined in any ratio in the same dosage unit. Combination products can be prepared from pellets containing actives that are incompatible or have different release profiles. Small pellets can be used as a method to limit drug migration for low-dose actives. The physical characteristics of the active ingredients and excipients can be modified to improve physical properties and downstream processing. As an example, a low-density, finely divided active can be pelletized to increase density, to improve flow, and to limit dusting [6]. Functional coatings can be applied easily and effectively. Dense multiparticulates disperse evenly within the gastrointestinal tract. They can be used to minimize variation caused by gastric emptying, to prolong gastrointestinal transit times [7,8], or to improve the tolerance of some compounds. Regardless of the application, care must be taken to achieve the required sphere properties.

Sphere properties can be altered to achieve the desired effects by varying the formulation or process variables. As an example, pellets to be used in controlled-release coating applications will likely require significantly different physical requirements than spherical granules for use in increasing gastric dissolution or other immediate-release applications. A product to be coated for controlled release should have a narrow size distribution, good sphericity, and smooth surface characteristics as well as low friability. Additionally, the process should be extremely reproducible and, batch-to-batch, should yield a uniform size distribution. Batch-to-batch fluctuations in size distribution can result in varying release profiles. Once coated, the sphere should have the desired release characteristics. If the coated pellets are to be compressed into tablets, the pellets, as well as the coating, will require sufficient strength to withstand the forces of compression. Upon the disintegration of the tablet, the individual pellets must retain their original release profile. Other properties such as flow, density, friability, porosity, and surface area are important properties for pellets or spherical granules intended for encapsulation or compression into tablets. Regardless of the dosage form, target release specifications must be met, and the release profile will undoubtedly depend on sphere properties.

A product produced using extrusion/spheronization can range from barely shaped, irregular particles with physical properties similar to a conventional granulation, to very spherical particles that are extremely uniform in size and shape [9]. Characteristics can be modified by altering the composition of the spherical particles [10], granulating fluid [11], or the process conditions used [12]. Compaction studies conducted on pellets similar to those used for controlled-release applications show that the bonding and the densification that occur during extrusion/spheronization can alter the deformation characteristics of some materials [11]. Microcrystalline cellulose (MCC), which deforms plastically in the dry powder state, exhibits an elastic deformation followed by a brittle fracture once extruded and spheronized under proper conditions [10]. For example, the amount of granulating fluid typically required for preparing reasonably sized pellets for controlled release results in relatively dense, larger-sized, spherical particles [13]. While these particles are ideal for controlled-release applications, the deformation characteristics, coupled with the larger size, result in reduced bonding sites and the production of weak compacts. These applications, controlled-release pellets and spherical granules for compression, are extreme examples requiring vastly different particle properties. The point is not to dwell on the properties required for each application, but rather to reinforce the fact that each application will have very specific requirements. One must first understand the properties required and then tailor the process to yield the desired effects. The effects of formulation and process variables will be discussed below.

A review of the literature shows that most investigators have tried to understand the small components of this process isolated from other effects. They have focused on particular formulation or process parameters or variables. While it is valuable to have a detailed understanding of the main variables, this approach fails to take into consideration the high degree of interaction that exists between the variables. The use of statistical experimental design is a valuable tool to understand not only the main effects, but also the interactions that can have a profound effect on the characteristics of the resulting pellets [14–17]. Additionally, these techniques are extremely useful during product/process development to understand the effect of variables and to control them to produce products having desired attributes. Factorial designs are useful for screening studies to determine the formulation and the process variables that have significant positive or negative effects [14,15]. Mixture design type studies have proven useful in formulation development. Designs that allow response surfaces to be modeled are useful in optimization [16,17]. Similarly, sequential simplex design methods are useful when a large number of variables need to be optimized. After pointing out the benefits of design methodology in this application, it should be understood that, for simplicity, much of the discussion to follow addresses the various topics individually. In reality, however, variables and their effects cannot truly be isolated from one another. Experimental design methods can be essential to the successful application of extrusion/spheronization. This chapter will review and discuss the general process, equipment types, and effects of formulation and process variables on the properties of spheres or pellets.

III. GENERAL PROCESS DESCRIPTION

Extrusion/spheronization is a process requiring at least five units of operation including dry mixing, wet granulation, extrusion, spheronization, and drying, with an optional sixth step—screening. For controlled-release applications, a coating step may also be necessary. First, the materials are dry mixed (1) to achieve a homogeneous powder dispersion and then wet granulated (2) to produce a sufficiently plastic wet mass. The wet mass is extruded (3) to form rod-shaped particles of uniform diameter that are charged into a spheronizer and rounded off (4) into spherical particles. The spherical particles are then dried (5) to achieve the desired moisture content and optionally screened (6) to achieve a targeted size distribution. A functional coating must be applied for controlled-release products regulated by a diffusion membrane. Each of the process steps is reviewed in greater detail below. The process flow diagram, shown in Figure 1, shows each of the process steps along with critical variables associated with them [18]. The end product from each of the steps is shown in Figure 2.

IV. EQUIPMENT DESCRIPTION AND PROCESS PARAMETERS

A. Dry Mixing

During the first step, powders are dry mixed to achieve a uniform blend prior to wet granulation. This is generally carried out in the same mixer used for the granulation. However, if a continuous granulator is used, a separate mixer is required for the dry mix. The dry mixing step is typically taken for granted because wet massing follows. The uniformity of the dry mix, however, can have a significant effect on the quality of the granulation and, in turn, the spherical particles produced. An uneven distribution of materials having wide differences in properties such as size and solubility can result in localized overwetting, at least initially, during the granulation step. The more soluble and finely divided components can also dissolve and become part of the

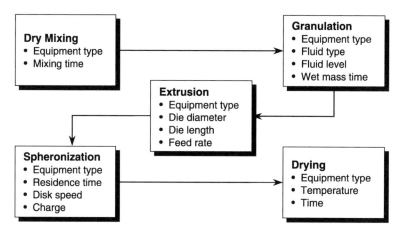

Figure 1 Process flow chart of the extrusion/spheronization process showing the process variables for each individual step. (From Ref. 14.)

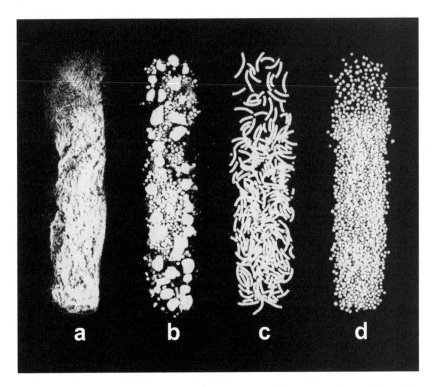

Figure 2 Product produced by the first four extrusion process steps: (a) powder from dry mixing, (b) granules from granulation, (c) extrudate from extrusion, and (d) spheres from spheronization.

granulating fluid. The fluids, rich in soluble compounds, can either remain as overwet regions or redistributed with continued wet massing [19]. If the overwet regions remain after wet massing, they can have adverse effects during the extrusion and spheronization steps. The unevenly wetted particle surfaces likely will result in pellets with a broad size distribution. Pellet uniformity (size and shape) is very much dependent on the uniform distribution and the composition of the granulating fluid, which includes not only the solvent but also any dissolved ingredients as well.

B. Granulation

The second step is granulation, during which a wet mass having the requisite plasticity or deformation characteristic is prepared. With a few exceptions, this step is similar to conventional granulation techniques used to produce product for compression. A batch type mixer/granulator is generally used; however, any equipment capable of producing a wet mass, including the continuous type, can be used. Batch-type processors include planetary mixers, vertical or horizontal high shear mixers, and sigma blade mixers. Examples of continuous mixers include the Nica M6 instant mixer [20] and high-shear, twin-screw mixer/extruders [21,22].

The major differences in the granulation step, as compared to typical granulations for compression, are the amount of granulating fluid required and the importance of achieving a uniform dispersion of the fluid. The amount of fluid needed to achieve pellets of uniform size and sphericity is likely to be greater than that required for a similar granulation intended for tableting. Likewise, the amount of mixing time required to achieve sufficient granulating fluid uniformity is likely to be greater than that used for a tableting granulation. Overgranulation and the resulting dense particle produced are typically not a concern in extrusion/spheronization, as it is in the production of a granulation intended for compression.

Instruments, such as a ram extruder [23] and a torque rheometer [24,25], have been used to characterize the deformation and flow characteristics of granulations used in extrusion/spheronization. They are useful tools in quantifying the rheological effect of formulation and the process variations in the granulation.

The ram extruder has been used as a tool to characterize the flow of wet masses through a die. The force profile produced as the mass flows through a die has been divided into three stages: (1) compression, (2) steady state flow, and (3) forced flow. The compression stage is where the materials are consolidated under slight pressure. Better formulations have a minimal compression stage with a rapid increase to a maximum force. The slope representing the force increase should be close to $90°$ [24]. Steady state flow is where the pressure required to maintain flow is constant. Low steady state forces or an evidence of moisture exiting the die prior to the extrudate generally indicates an overwet mass. An evidence of moisture at the die is often seen with a ram extruder but not with other extruder designs. High forces result in an extrudate with excessive surface irregularities. The irregularities are commonly referred to as sharkskin. While significant surface irregularities are undesirable, some irregularities can be beneficial. They assist in the extrudate breaking up into appropriate-sized, rod-shaped particles during the initial stages of the spheronization process.

Forced flow is where an increase in force is required to maintain flow. This condition is typically due to an insufficiently plastic mass. If the need for increased force happens early on without a significant evidence of a forced flow stage, it is likely due to an insufficient amount of granulating fluid. Granulating fluid at the extrudate and die interface acts as a lubricant and facilitates the movement of the extrudate through the die. The movement of fluid under pressure can cause a change from steady state to forced flow. This is a condition that should be minimized. Extrusion in a ram extruder is continuous, and this phenomenon is less likely to be seen in extruders that are not continuous, such as gravity-fed models [26]. These issues will be reviewed further during the discussion on extrusion. The three stages are shown in the force vs. displacement profile in Figure 3. A diagram of a ram extruder is shown in Figure 4.

The mixer torque rheometer (Caleva MTR) is a valuable tool for use in characterizing wet powder masses regardless of their intended application. Mean torque and torque range or amplitude data can be used to assess the stage of granulation and the degree of spreading or interaction. As indicated by Rowe and Parker, torque range (amplitude) will be at a maximum when the wet mass is at the funicular state of liquid saturation, when both voids and liquid bridges are present. The mean torque will be at a maximum when the wet mass is at the capillary state, when there is the largest number of capillary

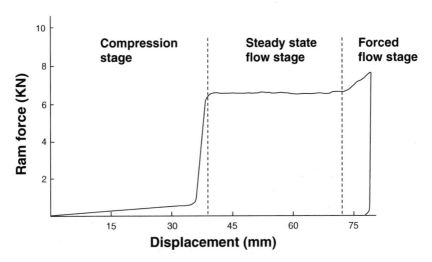

Figure 3 A force displacement profile for a microcrystalline cellulose/lactose/water mixture showing the three stages of extrusion on a ram extruder, compression, steady-state flow, and forced flow (ram speed 4 mm/s, die diameter 1.5 mm, R/L ratio 12. (From Ref. 24.)

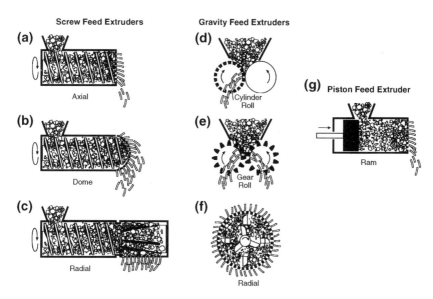

Figure 4   Schematic diagrams of extruder types used in extrusion/spheronization.

bridges between the granulating fluid and the substrate materials. The difference in the relative position of the two torque values (mean torque and torque
range) indicates the degree of binder/substrate interaction or spreading. A
greater difference between the two peak values indicates greater wetting and
spreading [25]. Some studies have indicated that better wetting leads to more
spherical, but larger, particles and smoother surfaces [20]. As stated previously, the fluid levels typically required to form a sufficiently plastic wet mass
for extrusion/spheronization are typically greater than those needed for a tablet
granulation. Tablet granulations are typically optimum at an advanced funicular state, but not as progressed as the capillary state. A wet mass at maximum
mean torque, or in the capillary state, is typically suitable for extrusion/
spheronization [24] (see Figure 5).

     Granulation variables such as granulating fluid level and wet mass
mixing time have a significant effect on the size of the pellets produced. At
relatively low granulating fluid levels, longer mixing times will distribute the
water more effectively and result in more cohesive granule surfaces. This
results in a slight increase in the size of the pellet produced. Conversely, when
higher fluid levels are used, longer mixing times will distribute the water
effectively within the pore structure of the particles, reducing or eliminating

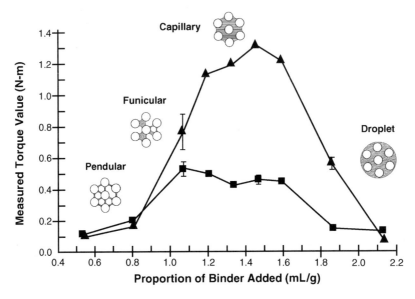

Figure 5 Variation in measure torque values with increasing binder content. (■) Amplitude of oscillation in torque; (▲) mean torque.

overwet surface while ensuring a sufficiently plastic mass. The distribution of water and the reduction of surface water result in a particle having a smaller mean size and narrower distribution [14]. Figure 6 shows the effect of granulating fluid, wet mass mixing time, and drug solubility of the mean size of pellets. Granulation times should be long, especially during the initial stages of development, prior to optimization. Granulating fluid levels will be discussed further during the review of formulation effects.

The use of batch-type mixers for granulation is well documented and understood. Having discussed the differences between a granulation intended for extrusion/spheronization as compared to compression, the equipment and the process will not be discussed further. There is value, however, in a brief discussion of continuous mixers. The instant mixer consists of a rapidly rotating disk or turbine onto which the powder mixture and the granulating fluid are continuously fed. The charge in the granulating zone or chamber, and therefore shear, can be controlled by adjusting the feed rate of the powder and fluid, or the gap of the granule outlet. The high-shear, twin-screw mixer/ extruders have screw feeders and mixer paddles and blades that are capable of shearing and kneading the feed materials. Dry powders and fluids are fed in

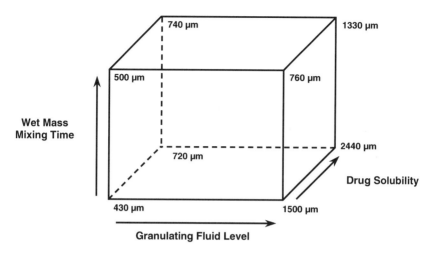

Figure 6   A box diagram showing the effect of wet mass mixing time, granulating fluid, and drug solubility. (From Ref. 14.)

through separate ports and mixed by the action of the screws paddles and blades as the materials progress through the mixing chamber or barrel. At the end of the chamber, a uniform wet mass is discharged. The mixer/extruder is capable of being configured to customize the amount of shear and energy used in the process by changing the configuration of the mixing paddles and blades. The configuration can have a significant impact on the properties of the granulation or extrudate produced [22]. As with the batch processors, it is critical to achieve a uniform level of fluid within the wet mass. The proper fluid/solids ratio is accomplished by maintaining a steady powder and fluid feed into the mixer/extruder. Both are critical; however, the powder feed is the most problematic. Small variations in feed rates can cause significant shifts in the moisture content of the granulation and therefore the quality of the spherical particles produced. For this reason, while continuous mixers can result in a more efficient process, especially for high-volume products, batch granulators are more widely used. An interesting advantage of a continuous mixer/extruder is that with slight modifications, such as the inclusion of a perforated plate at the discharge, the equipment can be used to accomplish two of the process steps—mixing and extrusion [27]. While this type of process is very difficult to control, it has been successfully applied.

Regardless of the mixer used, one must remember that the downstream process steps of extrusion and spheronization are very dependent on the level of granulating fluid contained in the granulation and the quality of its dispersion. High-energy mixers, such as high-shear mixers and high-shear, twin-screw mixer/extruders, can cause a significant rise in temperature. It may be necessary to use a jacket to guard against heat buildup. High temperatures can result in a greater than acceptable level of evaporation [28], or in an increase in the solubility of some of the solids. A reduction in fluid will reduce the plasticity of the granulation, while an increase in solubility will increase the weight ratio of granulating fluid because the solute is then part of that fluid [29].

C. Extrusion

The third step is the extrusion step, which forms the wet mass into rod-shaped particles. The wet mass is forced through dies and shaped into small cylindrical particles having a uniform diameter. The extrudate particles break at similar lengths under their own weight. The extrudate must have enough plasticity to deform but not so much as to adhere to other particles when collected or rolled in the spheronizer.

Extruders come in many varieties, but can generally be divided into three classes based on their feed mechanism. They include those that rely on a screw, gravity, or a piston to feed the wet mass into the extrusion zone [30]. Examples of extruders from each class are shown in Figure 4. Screw-fed extruders include the (a) axial or end plate, (b) dome, and (c) radial type, while gravity-fed extruders include the (d) cylinder, (e) gear, and (f) radial types. The screw- and gravity-fed types are used for development and manufacturing, with the radial varieties being the most popular for pharmaceutical applications. The piston-fed or ram extruder is primarily used in research as an analytical tool.

Screw extruders have either one (single) or two (twin) augers that transport the wet mass from the feed area to the extrusion zone. During the transport process, the screws compress the wet mass, removing most of the entrapped air. Studies have been conducted on the ram extruder to understand this compression or consolidation stage. They have shown that the apparent density of the wet mass plug prior to extrusion is approximately equal to the theoretical apparent particle density, indicating that nearly all of the voids were eliminated [31]. Twin-screw extruders generally have a higher throughput than single-screw models, while single-screw extruders compress and increase the density of the extrudate more. Other features that can affect the density of the extrudate are the spacing of the turnings on the screw and the space between the end of the screw and the beginning of the die [32]. Turnings that are wide and regularly spaced minimize the amount of compression during material transport. Screws with closer or progressively closer spacing between the turnings will result in more compression and produce a denser extrudate. The space between the screw and the die results in a void into which the material is deposited and compressed. The greater the space, the more compression takes place prior to extrusion. As the material builds up, the pressure increases and causes the material to be forced, under hydraulic pressure, to flow through the die. When the space between the screw and the die is at a minimum, extrusion takes place as the material is compressed in the nip, between the extruder blade and the die.

The primary difference between the various types of screw extruders is in the extrusion zone. An axial or dome extruder transports and extrudes the wet mass in the same plane. Axial extruders force the wet mass through a flat, perforated end plate, typically prepared by drilling holes in a plate. The thickness of the plate can be more than four times the hole diameter, resulting in high die length/radius (L/R) ratios. Longer dies result in a greater consolidation of the extrudate. Additionally, screens having low ratios are more fragile and susceptible to damage. The effect will be discussed further below. An LCI axial extruder is shown in Figure 7. Dome extruders use a dome or

Figure 7   An LCI axial endplate extruder. (Courtesy of LCI Corp.)

half-sphere-shaped screen as the die. The screen is prepared by punching holes in metal stock having a similar thickness as the hole diameter. This results in a die L/R ratio close to 2; however, variations in screen thickness are possible, resulting in a slightly higher or lower ratio. An LCI dome extruder is shown in Figure 8. Unlike axial and dome extruders, radial extruders extrude the wet mass perpendicular to the plane of transport. The material is transported to the extrusion zone where it is wiped against the screen die by an extrusion blade. The mass is forced through the die by the pressure generated at the nip. An LCI screw feed radial extruder is shown in Figure 9. As with dome-type extruders, the die is a stamped screen. Due to the shorter die lengths and the increased

Figure 8 An LCI dome extruder. (Courtesy of LCI Corp.)

Figure 9   Side view of the extrusion zone of an LCI screw feed radial extruder. (Courtesy of LCI Corp.)

number of holes or dies, dome and radial extruders have the advantage of higher throughput as compared to the axial type.

As with almost every step in extrusion/spheronization, heat buildup during extrusion is a significant concern. This is especially true of the screw-fed extruders. Axial extruders generate heat due to their long die lengths. Radial extruders can have a significant heat differential over the width of the screen. Materials fed into the extrusion zone will have the lowest temperature. However, as the material moves to the front of the zone, the temperature increases due to the longer residence time of the material. Of the screw-fed extruders, the dome type has the highest rates of throughput and is least likely to generate significant heat over an extended period. The short die lengths, however, result in a thin screen that deforms easily under pressure.

Gravity-fed extruder types include cylinder, gear, and radial designs. The cylinder and gear designs both belong to a broader class referred to as roll extruders. Both use two rollers to exert force on the wet mass and to form an extrudate. The cylinder extruder has rollers in the form of cylinders—one solid and one hollow with drilled holes to form the dies. The wet mass is fed by gravity into the nip area between the two cylinders and forced through the dies into the hollow of the cylinder. Gear-type extruders have rollers in the form of hollow gears. The dies are holes drilled at the base of each tooth. Wet mass is forced through the holes and collected in the hollow of the gears as the teeth and the base areas mesh. The last type of gravity-fed extruder to be discussed is the radial type. One or more arms rotate to stir the wet mass as it is fed by gravity. Rotating blades wipe the wet mass against the screen, creating localized forces sufficient to extrude at the nip. There is no compression prior to extrusion, which is the major difference between the gravity fed and the screw-fed radial extruders. This design results in very little heat effect. A Nica extruder is shown in Figure 10.

The primary extrusion process variables are the feed rate, die opening, and die length. The granulating fluid content of the granulation is also very critical because the properties of the extrudate and resulting pellets are very dependent on the plasticity, cohesiveness, and lubricity of the wet mass. The process variables and the fluid content have been the focus of many studies. Harrison et al. [23,31,33,34] studied the flow of the wet mass as it is forced through a die. They determined that steady state flow (described above and shown in Figure 3) was essential to produce a smooth extrudate that results in uniformly sized spherical particles having good sphericity and surface

(a)   **(b)**

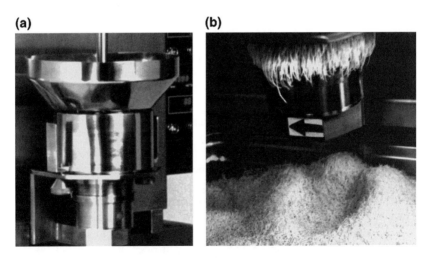

Figure 10   (a) Front view of a Nica gravity-feed radial extruder; and (b) close-up showing the extrusion zone. (Courtesy of Niro Inc.)

characteristics. Materials and processes that did not result in steady state, a condition referred to as forced flow, produced extrudates with surface impairments. In moderate cases, the surface is rough, while in more severe cases, a phenomenon commonly referred to as shark skinning occurs. With extruders other than a ram extruder, as mentioned previously, a small amount of shark skinning may be beneficial and may enable the extrudate to beak into appropriately sized lengths. In the absence of a ram extruder, a standardized syringe and an instrumented tension/compression tester (such as an Instron®; Instron Corp., USA) can be used [24]. Examples of a smooth extrudate and a shark-skinned extrudate are shown in Figure 11a and b, respectively.

Force displacement profiles of MCC and water at various ratios, MCC/lactose/water at a 5:5:6 ratio, and lactose/water at an 8:2 ratio, developed by Harrison et al., are shown in Figure 12. Steady state was possible with the MCC and MCC/lactose samples but not with lactose alone. As can be seen with the MCC samples, the duration of the compression stage was water level-dependent with no effect seen on the steady state stage. Additional studies indicated an effect of ram speed (extrusion speed) and die R/L ratio. An increase in ram speed increased the duration of the steady state stage with no effect on the compression stage. The R/L ratio had no effect on either compression or steady state. Wet mass composition therefore influenced the ability to achieve steady state, while the water level and ram speed influenced duration. Higher water levels decreased the force to produce steady state flow, but increased the duration. Faster ram speeds (extrusion rates) increased the duration of steady state and increased the force. As discussed below, other investigators have reported the correlation between extrusion force and sphere quality.

Harrison et al. also indicated that a uniform lubricating layer at the die wall interface must occur to eliminate the slipstick phenomenon responsible for forced flow. The development of a lubricating layer was dependent on the length of the die (a minimum length required), wall shear stress, and upstream pressure loss. They represent the frictional forces at the die wall interface and the estimated pressure loss at zero die length in the barrel of the ram extruder. The method for deriving these values is described in Ref. 18. These parameters allow for a quantitative comparison between formulations and process; however, no specific values can be targeted because they vary with materials.

Pinto et al. also showed that, at slow ram speeds, water moves toward the die wall interface and acts as a lubricant, resulting in reduced extrusion forces. At higher speeds, water is unable to move rapidly through the mass, resulting in higher forces [35]. They indicated that the water content and its distribution are critical in determining the particle size and the sphericity of the

(a)

(b)

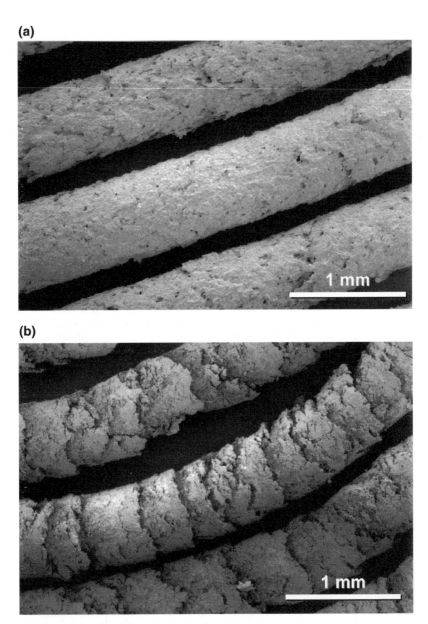

Figure 11 SEMs showing an example of a smooth extrudate (a) and an extrudate having surface impairment or shark skinning (b).

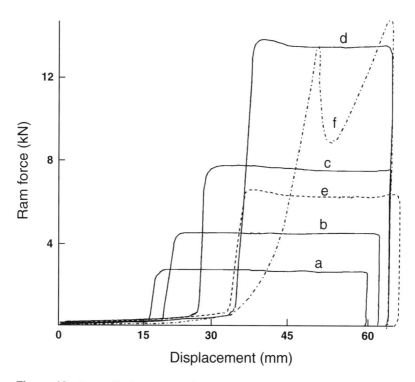

Figure 12 Force displacement profiles at various moisture contents of mixtures of microcrystalline cellulose and water: (a–d) microcrystalline cellulose/lactose/water (5:5:6) and (e) lactose/water (8:2) (f) at a ram speed of 4 mm/s, die diameter of 1.0, and an L/R ratio of 12. Moisture content of MCC/water mixtures (%): (a) 59.4, (b) 54.9, (c) 51.1, and (d) 45.0. (From Ref. 23.)

product. Lower water content and higher speed will reduce the size and the sphericity of the particles. The extrusion speed and the water content should be adjusted to achieve the desired effect. Other researchers have investigated the effect of die length using a gravity-fed radial extruder. Hellén et al. [36] indicated that the extrudate became smoother and more bound as the R/L ratio of the die was increased. Vervaet et al. [37] reported that a higher R/L ratio enables the use of lower water levels to achieve a more bound extrudate. This also increased the range (drug loading and water level) over which quality pellets could be produced. They attributed the increased latitude and capability to an increased densification and a resulting well-bound extrudate. The average pore diameter and the bulk density reported for extrudate prepared from various MCC/dicalcium phosphate/water ratios at two R/L ratios are shown in Table 1. Baert et al. [38] also indicated a similar increase in latitude when a cylinder extruder having an R/L ratio of 4 was compared with a twin-screw extruder having an R/L ratio close to 1.8.

Three types of screen perforations are possible. One method of producing a screen is by punching the holes, which is generally done for relatively thin screens. Another method is by drilling the holes, which is generally done for relatively thick screens. Additionally, screens having either straight or tapered die walls can be produced. Vervaet and Remon reported the effect of extrusion screen perforation. Higher drug loading was possible with the drilled or tapered screens as compared to the punched screen [39]. This was attributed to a variation in die filling, due to the irregularity of the punch holes and the additional compression that occurs in the die with the tapered hole.

Other studies have shown that there is an optimal pressure range over which an extrudate capable of yielding acceptable pellets can be produced. Shah et al. [40] demonstrated the correlation between screen pressure yield and density. A high yield of pellets within a narrow, targeted size distribution was produced as long as the screen pressure was maintained within a given range. The relationship between yield and screen pressure is shown in Figure 13.

While many of the researchers have indicated a need for a more cohesive extrudate, few have expressed a need to remove all surface impairments. Some researchers have indicated that pellets having acceptable characteristics can be produced from extrudates having shark skinning. O'Connor and Schwartz [41] have found the presence of shark skinning to be advantageous in facilitating the breakage of the extrudate during the spheronization step.

Experimental design studies conducted to concurrently investigate the effect of extrusion as well as other formulation and process variables have

Table 1 Average Pore Diameter and Bulk Density of Extrudate Composed of DCP/Avicel PH-101/Water Mixtures, Extruded Using Screens with a Different R/L Ratio (From Ref. 32.)

Composition DCP/Avicel/water (wt/wt)	L/R ratio of screen	Average pore diameter (µm)	Bulk density (g/ml)
150:380:470	4	0.982	1.132
150:400:450	4	0.992	1.211
150:380:470	2	1.249	0.949
150:400:450	2	1.292	0.947

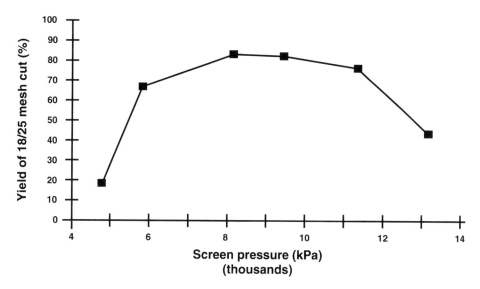

Figure 13   The effect of extruder screen pressure on the yield of particles within an acceptable distribution. (From Ref. 34.)

indicated the extrusion variables to be less significant than granulating fluid levels or variables of the spheronization step. Hasznos et al. [42] determined that extruder speed had little effect on the size distribution of the final product, or moisture change during processing as compared to the spheronization variables. Hileman et al. [43] indicated that when water/MCC ratios are held constant, a change in screen size results in a significant change in the size distribution. However, in a study where water level was included as a variable, Erkoboni et al. have shown that the effect of screen size on size distribution is small compared to the effect of a change in water level. A change in water level can shift the mean size and still result in an acceptable distribution [14]. This is in agreement with an earlier work by Jalal et al. [6], who also showed that the mean particle size is typically smaller than the size of the screen itself, due to the shrinking of the spherical particles during the drying step.

Thoma and Ziegler investigated the quality of pellets produced using three distinctly different extruder designs, a gravity-fed roll extruder, an axial, single-screw extruder, and a gravity-fed basket extruder. They showed that the granulating fluid level required to produce similarly sized pellets was different for each of the extruders. The single-screw extruder required the greatest amount of water, followed by the gravity-fed roll extruder, which required less, and the gravity-fed basket extruder, which required the least amount of water. The axial, twin-screw extruder had the widest granulating fluid range over which pellets of the same particle size could be produced [44]. While formulation-dependent, the differences in granulating requirement can be from moisture loss due to heat buildup [45].

D. Spheronization

The fourth step in the extrusion/spheronization process is the spheronization step. It is carried out in a relatively simple piece of equipment. The working parts consist of a bowl having fixed sidewalls with a rapidly rotating bottom plate or disk. The rounding of the extrudate into spheres or pellets is dependent on frictional forces. The forces are generated by particle–particle and particle–equipment interactions. For this reason, the disk is generally machined to have a grooved surface that increases the forces generated as particles move across its surface. Disks having two geometrical patterns are typically produced—a cross-hatched pattern with the groves running at right angles to one another, and a radial pattern with the groves running radially from the center. The two varieties are shown graphically in Figure 14. Some studies have shown the rate of spheronization to be faster with the radial pattern; however, both plates will result in an acceptable product [30].

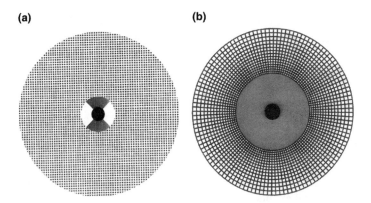

(a)   **(b)**

Figure 14 Spheronizer disks having two geometric patterns: (a) a cross-hatched pattern with the groves running at right angles to one another, and (b) a radial pattern with the groves running radially from the center.

During the spheronization step, an extrudate is transformed from rod-shaped pellets into spherical particles. This transition occurs in various stages. Once charged into the spheronizer, the extrudate is drawn to the walls of the extruder due to centrifugal forces. From here, what happens is very much dependent on the properties of the extrudate. Under ideal conditions, the extrudate breaks into smaller, more uniform pieces. Within a short period of time, the length of each piece is approximately equal to the diameter the attrition and the rapid movement of the bottom plate or disk. The differential in particle velocity as the pieces move outward to the walls, begin to climb the walls, and fall back onto the rotating bed—along with the angular motion of the disk—results in a ropelike formation [3]. A graphical representation of this ropelike formation is shown in Figure 15. This formation can be a critical indicator of the quality of the granulation or extrudate. As pointed out by Reynolds [3], the disk rotating without a movement of the product indicates an overwet condition. The condition is caused either from a granulation that was initially overwet, or from the migration of water or a fluid ingredient to the surface of the extrudate during extrusion or spheronization. One way to determine the appropriate range for granulating fluid is to prepare successive batches, each with increasing fluids levels. If a fluid level results in a static

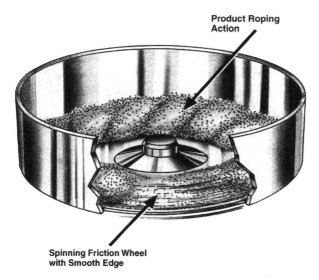

Product Roping Action

Spinning Friction Wheel with Smooth Edge

Figure 15 A graphic representation of the characteristic ropelike formation in a spheronizer bowl during operation.

bed, while the spheronizer is running, the fluid level is too high and the next lower level is the maximum fluid that can be employed.

As mentioned, the transformation from cylinder-shaped extrudate to a sphere occurs in various stages. Two models have been proposed to describe the mechanism and are shown graphically in Figure 16. The model proposed by Rowe [30] describes a transition whereby the cylindrical particles (Figure 16-2a) are first rounded off into cylindrical particles with rounded edges (Figure 16-2b), then form dumbbell-shaped particles (Figure 16-2c), ellipsoids (Figure 16-2d), and finally spheres (Figure 16-2e). The second model proposed by Baert et al. [38] suggests that the initial cylindrical particles (Figure 16-1a) are deformed into a bent rope-shaped particle (Figure 16-1b), and then form a dumbbell with a twisted middle (Figure 16-1c). The twisting action eventually causes the dumbbell to break into two spherical particles with a flat side having a hollow cavity (Figure 16-1d). Continued action in the spheronizer causes the particles to round off into spheres (Figure 16-1e). When the sphere is fractured, a hollow particle is revealed [46]. The

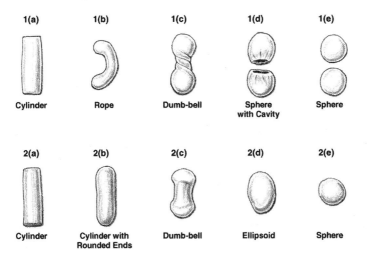

1(a)	1(b)	1(c)	1(d)	1(e)
Cylinder	Rope	Dumb-bell	Sphere with Cavity	Sphere

2(a)	2(b)	2(c)	2(d)	2(e)
Cylinder	Cylinder with Rounded Ends	Dumb-bell	Ellipsoid	Sphere

Figure 16 A graphic representation of the two models proposed to describe the mechanism of spheronization. The model proposed by Rowe [30] describes a transition from cylindrical particles (2a) into cylindrical particles with rounded edges (2b), dumbbells (2c), ellipsoids (2d), and spheres (2e). The model proposed by Baert et al. [38] describes a transition from initial cylindrical particles (1a) into a bent rope (1b), dumbbell (1c), two spherical particles with a hollow cavity (1d), and spheres (1e). (From Refs. 24,37.)

exact mechanism is likely composition-dependent. If the extrudate is overwet, particle growth will occur, resulting in a broad size distribution. Underwet extrudate will not have enough plasticity to further round off in the spheronizer; the result is the formation of dumbbells. The scanning electron microscopes (SEMs) in Figure 17 show an example of good spheres or pellets produced from a sufficiently plastic mass and dumbbells that would not deform further.

Variables in the spheronization step include spheronizer size, charge, disk speed, and residence time. A number of studies have shown that each of

(a)

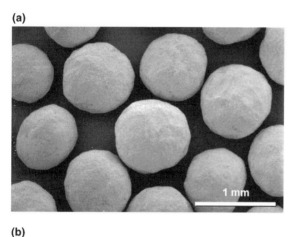

(b)

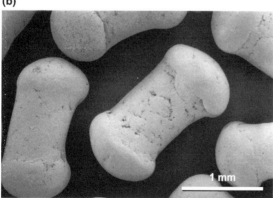

Figure 17   An example of good spheres produced from a sufficiently plastic mass (a) and dumbbells that would not deform further produced from underwet extrudate (b).

the variables has the potential to play a significant role in influencing the physical characteristics of the resulting product. Hasznos et al. showed that a higher disk speed and a longer residence time increased the coarse fraction and the mean diameter and decreased the fine fraction. The faster speed and longer time also increased the moisture loss during the process. Because the moisture loss can reduce the plasticity of the particle, it can have the same effect as an underwet granulation. The particles may not round off into pellets and stay as deformed cylinders or dumbbells. Higher spheronizer charges reduced the moisture loss. They also suggested that an interaction between spheronizer speed and residence time indicated that the total number of revolutions of the disk was critical. A change in one of the variables could be offset by an opposite change of the other, as long as the total number of revolutions remained constant [42]. Hellén et al. showed a similar moisture loss during spheronization. In addition, they indicated that the major factors influencing the shape of the pellets were disk speed and residence time. High speed and long time produced more spherical particles [20]. Wan et al. indicated that a minimum disk speed and residence time were required to round the cylinder-shaped extrudate. Furthermore, an increase in speed or time, up to a limit, increased the median diameter of the pellets while higher speeds and longer times caused a reduction in size. Short residence times at high disk speeds resulted in small but round particles [47]. Similar results, with regard to speed, were reported by Newton et al. They found that at too low a speed, there were no significant changes in the shape of the extrudate while too high a speed resulted in a reduction in size [48].

A number of investigators have reported the effect of disk speed and residence time on density. Woodruff and Nuessle [9] reported the variables to have no effect on the density of the pellets as compared with the density of the granulation and the extrudate. These results conflict with most of the other studies; however, they are likely due to the use of mineral oil in the formulation. The oil can reduce the frictional forces at the die wall during extrusion, and between particles and equipment surfaces during spheronization. A number of investigators, including Malinowski and Smith, reported that an increase in either disk speed or residence time resulted in an increase in density [13,14,20,43]. This is in agreement with a study conducted by Bataille et al. [49] that indicated an increase in disk speed resulted in a decrease in porosity. In addition, they showed that an increase in disk speed resulted in an increase in sphere hardness.

O'Connor et al. [18] indicated that the friability of placebo pellets decreased with increasing residence time while the mean particle diameter decreased. Hellén and Yliruusi [50] studied the effect of residence time and disk speed on pellet shape. A faster speed and a longer residence time resulted in rounder pellets.

E. Drying

Drying is the final step in the process, excluding optional steps such as coating and encapsulation or tableting. This can be accomplished in any dryer that can be used for conventional-type granulations, including tray dryers, column-type fluid beds, and deck-type vibratory fluid beds. Each of the drying techniques has advantages; however, the major differences are based on the rate of fluid removal. Tray drying is the slowest of the processes. Fluidized bed dryers result in a much more rapid drying rate because of the higher air volumes and the potential use of higher inlet temperatures. Column fluid beds are batch dryers, while the deck-type dryers offer the advantage of a continuous process. Both have been used successfully in drying products produced by extrusion/ spheronization. The drying process must be chosen based on the desired particle properties.

Because tray drying is a slow process in a static bed, it can offer the greatest opportunity for a drug to migrate toward the surface and to recrystallize [51]. The more rapid rate in a fluid bed will likely minimize the effects of migration. This phenomenon can have an effect on a number of particle properties. The increased active concentration at the surface of the particle can increase the rate of dissolution. This recrystallization, however, can cause a problem for applications requiring film coating because the smooth surfaces developed by the spheronization process would be damaged. Additionally, the crushing strength of tray-dried particles will likely be greater than their fluid bed counterparts. The slow recrystallization in the static bed allows for crystal bridges to develop as the fluid is removed and the solute recrystallizes.

V. FORMULATION VARIABLES

The composition of the wet mass is critical in determining the properties of the particles produced. This is clearly understood if we look at what material behaviors are required during each of the process steps. During the granulation step, a plastic mass is produced—a simple enough task if ended there. The materials must form a sufficiently plastic mass, deform when extruded, and break off to form uniformly sized cylindrical particles. A minimal amount of granulating fluid should migrate to the surface during extrusion, and the particles should stay discrete during collection. During spheronization, the particles must round off to form uniformly sized pellets. They must not dry out because of temperature or air volume, or grow in size as a result agglomeration. The fact is, many are asked from materials used in this process. This is especially true of formulations containing high percentages of actives, where low levels of excipients are used to impart the desired properties to the mass.

The importance of using sphere-forming excipients was noted early on. Conine and Hadley [2] cited the necessity of using MCC. Reynolds [3] went on to indicate the need for either adhesives or capillary-type binders. He cited cellulose gums, natural gums, and synthetic polymers as adhesives, and MCC, talc, and kaolin as capillary-type binders. Since then, much work has been conducted in an attempt to understand the significance of material properties. Some of the studies are discussed below.

O'Connor et al. studied the behavior of some common excipients in extrusion/spheronization. The materials were studied as single components using water as the granulating fluid in an attempt to understand their application in the process. Of the materials tested, only the MCC or MCC with Na-CMC were capable of being processed. Other excipients, including dicalcium phosphate, lactose, starch, and modified starch, did not process adequately [18]. This is in agreement with other studies, which showed that most could be used as coexcipients but did not perform well on their own.

In an additional study, O'Connor et al. investigated the effect of varying drug/excipient ratios. At low drug levels, they found that the spheronizing excipient played the most significant role in determining sphere properties. They found that, for low-dose applications, MCC was the best excipient to use because it formed the most spherical particles. At moderate drug loading (50%), MCC, as well as the two products consisting of MCC coprocessed with Na-CMC (Avicel® RC-581 and Avicel CL-611), resulted in acceptable pellets. At higher loading levels, however, the MCC did not yield acceptable pellets and the coprocessed materials did. The pellets produced using Avicel CL-611 were the most spherical. In addition, they found dissolution to be dependent on the type of excipient used, the solubility, and the concentration of the actives. Pellets containing MCC remained intact and behaved as inert matrix systems, while those containing the coprocessed products formed a gel plug in the dissolution basket and were described as water-swellable hydrogel matrix systems. The release profiles for pellets containing each of the excipients and theophylline in a 50:50 ratio are shown in Figure 18. The release profiles for pellets containing different drug loads are shown in Figure 19. An increase in drug load resulted in an increased release rate. The release profiles for pellets containing actives having different solubilities, including chlorpheniramine maleate, quinidine sulfate, theophylline, and hydrochloro-

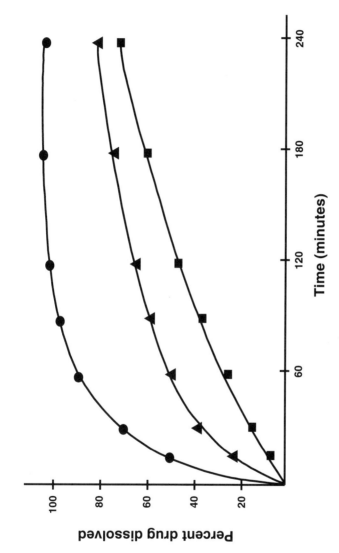

Figure 18 Dissolution profiles of spheres containing 50% theophylline in different Avicel MCC types. (●) Avicel PH-101, (▲) Avicel RC-581, (■) Avicel CL-611. (From Ref. 8.)

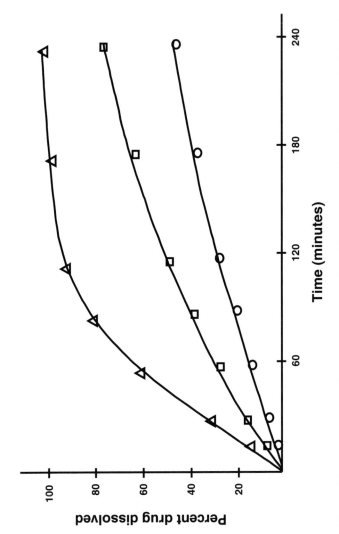

Figure 19 Dissolution profiles of spheres containing different concentrations of drug in Avicel CL-611. (○) 10%, (□) 50%, (△) 80%. (From Ref. 8.)

thiazide, are shown in Figure 20. An increase in drug solubility resulted in an increased release rate [52].

Millili and Schwartz demonstrated the effect of granulating with water and ethanol at various ratios. The physical properties of the pellets changed significantly as the ratio of the two fluids was varied. Pellets could not be formed with absolute ethanol, but were possible with 5:95 water/ethanol. An increase in the water fraction resulted in a decrease in porosity, friability, dissolution, and compressibility, and an increase in density. The porosity of pellets granulated with 95% ethanol was 54% while the water-granulated product had a porosity of 14%. When greater than 30% water was used, pellets remained intact throughout the dissolution test. As previously discussed, water-granulated pellets were very difficult to compress, while pellets granulated with 95% ethanol were significantly more compressible than those prepared using water [11]. Tablet hardness vs. compression forces profile is shown in Figure 21. In a later study, Millili et al. proposed a bonding mechanism, referred to as autohesion, to explain the differences in the properties of pellets granulated with water and with ethanol. Autohesion is a term used to describe the strong bonds formed by the interdiffusion of free polymer chain ends across particle–particle interfaces [53].

Using a ram extruder, Harrison et al. demonstrated that steady state flow could not be achieved with lactose. Additionally, they demonstrated the reduced sensitivity of MCC to small changes in moisture as determined by the force required to induce plug flow in a cylinder. Comparing MCC to a MCC/lactose blend and 100% lactose, they found that, with lactose, small changes in moisture caused large changes in force, while with MCC, larger changes in moisture were required to have similar effects on the force [23].

Baert et al. used mixtures of MCC and coexcipients at various ratios to demonstrate the effect of solubility and the total fluid on extrusion forces. They showed that if the coexcipient was insoluble, such as dicalcium phosphate, the force required for extrusion increased with increasing levels of coexcipient. When a soluble excipient such as lactose was used, the force required to extrude decreased with the addition of the initial amounts of lactose. After a certain level, however, the reduction in force stopped and the force began to increase. This was due to the initial solubilization of lactose and the resulting increase in the total fluid level. Once the fluid was saturated, the remaining lactose was not soluble and the force began to increase. The increase began at about 10% lactose level for α-lactose and 20% for β-lactose. This was due to the difference in solubility between the two lactose materials [28]. The effects of dicalcium phosphate and various lactose grades on extrusion force are shown in Figure 22. Similarly, a study to determine the

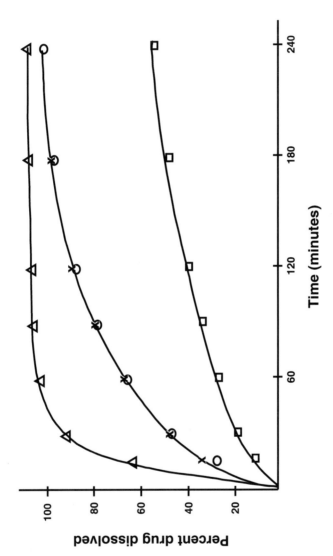

Figure 20 Dissolution profiles of spheres containing 10% drug in Avicel PH-101. (△) Chlorpheniramine maleate, (○) quinidine sulfate, (✕) theophylline, and (□) hydrochlorothiazide. (From Ref. 8.)

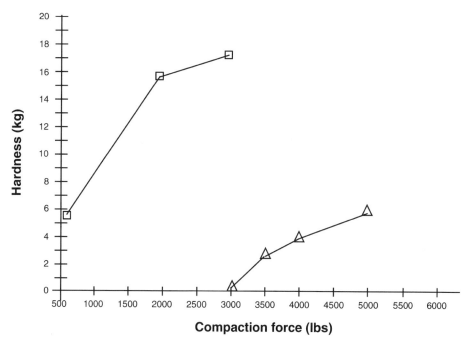

Figure 21   The effect of varying compression force on hardness of compacted 16/30 mesh spheres of 10% theophylline/Avicel PH-101. (△) Spheres prepared by water; (□) spheres prepared by 95% ethanol granulation. (From Ref. 9.)

influence of drug solubility was conducted by Lustig-Gustafsson et al. using equal parts of model drugs and MCC. The granulating fluid level required was dependent on the model drug solubility and its particle size. The optimum quantity of water was found to decrease linearly as a function of the natural log of the drug solubility [54] (see Figure 23).

Newton et al. showed that excipient source could have an effect on the quality of pellets as determined by size and sphericity. Various sources of MCC were evaluated. At fixed granulating fluid levels, significant differences in percent yield were seen among the materials within given size ranges. Some of the differences were greater than 50% [55].

Funck et al. showed that low levels of common binders could be used to produce high drug-loaded pellets with MCC. Materials such as carbomer, sodium carboxymethylcellulose (Na-CMC), hydroxypropylcellulose (HPC), hydroxypropylmethylcellulose (HPMC), povidone (PVP), and pregelatinized

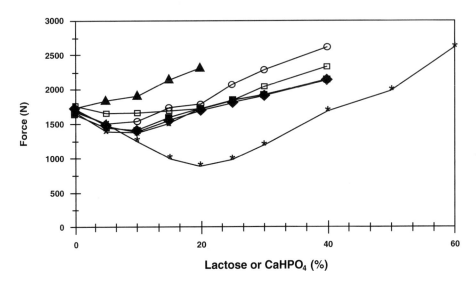

Figure 22 Influence of the amount of lactose or dicalcium phosphate dihydrate (percent total weight) on the extrusion forces (N) for mixtures of lactose or dicalcium phosphate dihydrate/Avicel PH 101/water after granulation with a planetary mixer. Each point is the mean of six values. The SD is lower than 3% for each point. Six different types of lactose were used: α-lactose monohydrate 80 mesh (□), α-lactose monohydrate 200 mesh (○), α-lactose monohydrate 325 mesh (♦), spray dried lactose DCL 11 (■), anhydrous β-lactose DCL 21 (×), and anhydrous α-lactose DCL 30 (⋆); and one type of dicalcium phosphate dihydrate was used (▲). (From Ref. 22.)

starch were used. All materials were capable of producing pellets of acceptable quality. Dissolution testing showed that pellets containing HPC and HPMC remained intact during testing, while pellets containing starch, PVP, and Na-CMC disintegrated [56].

Linder and Kleinebudde reported that pellets produced with powdered cellulose had a higher porosity and a faster dissolution than those made using MCC. Pellets could not be produced using only powdered cellulose and drug; a binder was required. The higher porosity of the pellets prepared from powdered cellulose may be beneficial for applications requiring compression [57].

Fielden et al. showed that increasing the particle size of lactose resulted in forced flow and high extrusion forces, which resulted in poor-quality extrudate and pellets having a wide size distribution. This was attributed to the increased pore diameter of the mixture containing the coarse lactose that allowed a greater movement of water [58].

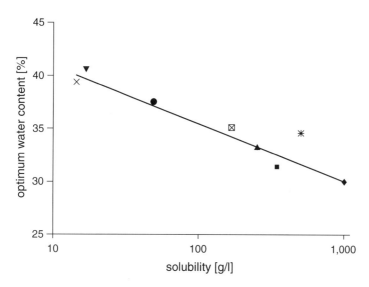

Figure 23 Optimum water content as a function of the natural logarithm of the solubility of the model drugs: (✳) 5-aminosalicylic acid; (■) ascorbic acid; (●) boric acid; (▼) caffeine; (◆) glucose; (▲) glycine; (⊠) lactose monohydrate (fine); (×) paracetamol.

Chien and Nuessle showed that the use of a surfactant, such as sodium lauryl sulfate (SLS), reduced the migration of drugs to the surface of the sphere during drying by reducing the surface tension of the granulating fluid. The reduction in surface tension also made it difficult to produce a cohesive extrudate in some cases. Mesiha and Vallés found that surfactants, such as SLS and Polysorbate 80, resulted in extrudate and pellets having very smooth surfaces, while common tablet lubricants, such as magnesium stearate, formed extrudate with severe shark skinning and pellets with poor surface characteristics [59].

Some miscellaneous observations include the following. Reynolds [3] reported that excess extrudate friability could be overcome by incorporating more MCC, binder, or water in the granulation. Erkoboni et al. [14] indicated that sphere hardness was most affected by the level of MCC in the formulation and the level of granulating fluid used. In addition, Erkoboni et al. showed that an increase in extruder screen size resulted in reduced friability. Hileman et al. [43] showed that MCC had a narrower water range over which quality pellets could be made than MCC coprocessed Na-CMC. Hellén et al. [20] showed that the surface characteristics were influenced by the water level, with higher water levels giving smoother surfaces. Law and Deast determined that coprocessed extrusion aids prepared from MCC and hydrophilic polymers are more functional than the same materials in a physical mixture. The characterization of the material revealed that the polymer allowed the MCC to disintegrate into smaller crystallites upon wetting [60].

Schwartz et al. demonstrated that the compaction characteristics of MCC processed into pellets are significantly different than the original powder. The powder material forms hard compacts at low compression forces, while the pellets are not compressible and form soft compacts, even at high forces. They indicated that pellets prepared from MCC showed a high degree of viscoelasticity over the entire compression range. The inclusion of coexcipients, such as lactose and dicalcium phosphate, increases the compactability by decreasing the viscoelastic resistance or pressure range over which the pellets behave elastically. A reduction in viscoelastic resistance was seen with pellets containing both lactose and dicalcium phosphate; however, dicalcium phosphate had a greater effect. Compaction profiles of pellets containing 10% theophylline with either MCC, MCC/DCP, or MCC/lactose in a 22.5:67.5 ratio are shown in Figure 24. This is important both for the preparation of drug-loaded pellets and placebo spherical granules. Drug-loaded pellets should be formulated to be very robust. If the pellets are to be compressed into tablets, they should be capable of withstanding the force of compression. Conversely, placebo spherical granules can be used as compressible beads in the compression of drug-loaded pellets into tablets. They should be capable of deforming plastically at significantly lower forces than required to rupture the drug-loaded pellets.

Many of the functional materials used in extrusion/spheronization contract or shrink during drying of the pellets. Kleinbudde [61] discusses this effect and the results of a study with MCC and low substituted HPC. Both materials uptake water readily. Greater water uptake during wet processing results in a greater contraction during drying. This effect was partially reversible and some degree of recovery was seen when the dried pellets are exposed to water [62]. Other materials that display this include croscarmellose and cross-linked polyvinylpyrrolidone.

There are a number of useful applications for this swelling effect. Smaller and/or denser pellets having lower porosity can be prepared. This can result in better packing density for capsule filling. It can also affect dissolution by reducing the porosity and therefore the channels for drug migration to the surface of rehydrated pellets. Spherical particles, to be included in immediate-release tablets, can be made fine enough to blend effectively and compress. One of the most established applications for pellets produced by extrusion/spheronizations is multiparticulates for controlled release. The effect of contracted

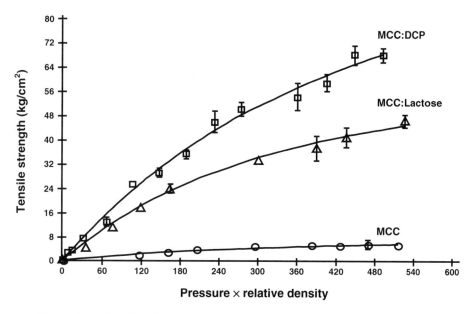

Figure 24   The effect of excipients on the compaction profile of spheres. Compaction profiles spheres containing 10% theophylline with either MCC, MCC/DCP, or MCC/lactose in a 22.5:67.5 ratio using the Leuenberger model. (From Ref. 7.)

particles that expand during hydration can have significant negative consequences if not recognized early on and addressed. Once coated and submersed in fluid, the swelling pellet can put so much pressure on a film coat, causing it to rupture and resulting in rapid release. This must be taken into consideration when developing a pellet and coating system for use in controlled-release applications. The objective should be to minimize the amount of swellable material and to maximize the strength of the film to be applied.

VI.   SCALE AND SCALE-UP

As with any development effort, the initial scale chosen for the development of an extrusion/spheronization process and formulation is likely based on the availability and the cost of the active. Equipment manufactures have designed small benchtop-scale extruders and spheronizers to accommodate the need for preparing gram-size batches. The mixer torque rheometer, manufactured by Caleva, can be effectively used as a granulator in addition to characterizing the granulation. Rowe and Iles [63] have reported on the use of the Caleva spheronizer. Wan et al. [64] have demonstrated the use of a bench-scale spheronizer using a batch weight of less than 50 g.

Extruder settings are not likely critical at this stage as the conditions used to produce an extrudate at this scale will not translate to other equipment. The objective should only be to produce an extrudate that can be spheronized. Another option is to use a ram-type extruder and to characterize the wet mass as it is being extruded. Spheronizer variables include charge, residence time, and disk speed. The charge should be as large as possible. The residence time will likely be longer than at a larger scale. Depending on the formula, 15 min may not be unreasonable. Disk speed can be very fast due to the small diameter of the disk. Values exceeding 1000 rpm are typical. As with many processes involving rotational parts, it is not the speed of the disk that influences the transformation of the extrudate from a rod-shaped particle to a pellet, but rather the peripheral tip speed. With the high speed of the disk and prolonged residence times, care should be taken to minimize the fluid evaporation.

The scale-up of an extrusion/spheronization process essentially involves the separate scale-up of each of the five individual process steps: dry mixing, granulation, extrusion, spheronization, and drying. This discussion will primarily focus on the process steps unique to extrusion/spheronization. The scale-up of the other steps is well documented in the literature.

Aside from the benchtop extruders, most of the development works on extrusion are likely conducted on an extruder very similar to what will be used in production. The primary difference among laboratory development, pilot scale, and production will likely be the batch size. Because extrusion is a continuous process, the primary change, as the scale increases, is the duration of the run. A number of considerations should be taken. As the run duration increases, the temperature has a better chance to come to equilibrium. If equilibrium represents a higher operating temperature than with smaller batches, evaporation may occur. Evaporation should be controlled, or the fluid level of the granulation should be slightly increased. The better option is to control evaporation. Depending on the design of the extruder, a cooling jacket may be an option. Another consideration is the longer holdover time of the granulation. Longer extruder run times mean larger granulation batches. Changes in the granulation can occur as it sits, especially if there are soluble components. As the material sits or as the temperature increases, a larger quantity of the soluble components can go into the solution, increasing the effective granulating fluid level. As was discussed earlier, this increased fluid level can result in a larger size distribution of pellets. To best understand this effect, holdover studies should be conducted on smaller batches to determine the effect of processing materials that have "aged."

Spheronization is a primary concern during scale-up. As mentioned previously, process variables include disk speed, residence time, and charge. Spheronization is a semicontinuous process. Discrete batches are spheronized consecutively. Each of the spheronizer charges is only a fraction of the total batch. Two potential effects should be closely monitored. Repetitive charges in the spheronizer can cause heat buildup. In turn, this can induce sticking on the disk, or the production of larger particles because of an increased solubility of some materials. An increase in the volume of air delivered to the gap between the spheronizer wall and disk can help offset some of the effects of heat. Another product of scale-up are the likely effects because of the increased mass or charge in the spheronizer. For an average laboratory-scale, pilot-scale, and small production-scale spheronizer, a full charge is likely between 4 and 6 kg of wet extrudate. Laboratory experiments can be carried out on the same equipment at a charge of about 1 kg. This fourfold to sixfold increase in mass

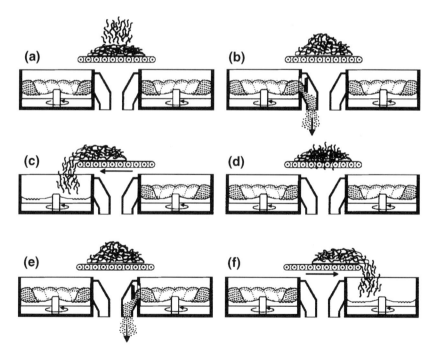

Figure 25 A graphic representation of a twin spheronizer shuttle system using two spheronizers in parallel and a shuttle receptacle. When both units are full, the shuttle receptacle collects the extrudate (a). After one empties (b), the shuttle box fills it (c). The cycle repeats itself for the second unit (d–f). (From Ref. 26.)

can have a significant effect. Wan et al. [64] showed that an increased charge or residence time resulted in a reduced pellet size.

Of the two process steps unique to extrusion/spheronization, extrusion is a continuous process while the second, spheronization, is a batch process. To make the process viable for commercial operations, two systems have been developed to enable the extruder to continuously feed material to the spheronizer(s). The first system is a semicontinuous shuttle system, and the second is a cascade system. The shuttle system is typically used when uniform particles are required, such as for controlled-release coating applications. The cascade system, however, can be used for applications where less size and shape uniformity is required, such as granulations intended for compression.

The shuttle system uses two spheronizers in parallel. It is designed to fill one spheronizer while the second is in the middle of its cycle, to continue to collect extrudate in a shuttle receptacle while they are both full and operational, and to fill the second after it empties and the first unit is in the middle of its cycle. The shuttle system operation is shown graphically in Figure 25. A picture of a Caleva spheronizing system having twin spheronizers is also shown in Figure 26. The cascade operation uses one or more spheronizers that are modified to have the disks some distance below the discharge chute [32].

Figure 26 A Caleva spheronizing system having twin spheronizers and a shuttle receptacle. (Courtesy of GEI Processing Inc.)

This results in a spheronization zone having a fixed volume. A product is continually fed from either the extruder or a previous spheronizer. As the charge volume grows from the incoming material, some product is discharged. The residence time is dictated by the feed rate. The reduced size and shape distribution are because of the percentage of material that does not reside in the spheronization zone for the intended period of time. The number of spheronizers placed in sequence depends on the desired outcome. However, if only a slight rounding with minimal densification is required, one spheronizer with a short residence time will be sufficient. The cascade operation is shown graphically in Figure 27.

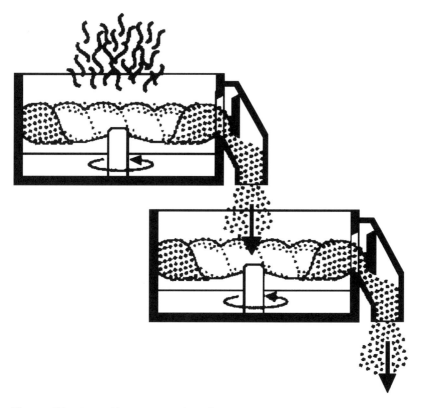

Figure 27 A graphic representation of two spheronizers in sequence to form a cascade system. The product is continually fed and as the charge volume grows, the product is discharged. (From Ref. 26.)

VII. SUMMARY

Extrusion/spheronization is a versatile process capable of producing pellets and, in some cases, spherical pellets having unique physical properties. Potential applications are many, including both immediate release and controlled release. Regardless of the application, care must be taken to understand the desired properties and the formulation and process variables capable of achieving them. Lastly, the use of a statistical experimental design for formulation and process development is strongly recommended because of the high degree of interactions between the variables.

ACKNOWLEDGMENTS

Special thanks to my family for their patience, Ron Vladyka for preparing samples used in the illustrations as well as review and comment, Lynn DiMemmo for preparing the SEMs, Chris Sweriduk for review and comment, and Lois McLean for administrative support.

REFERENCES

1. Nakahara. U.S. Patent 3,277,520, October 1966.
2. Conine JW, Hadley HR. Preparation of small solid pharmaceutical spheres. Drug Cosmet Ind 1970; 106:38–41.
3. Reynolds AD. A new technique for the production of spherical particles. Manuf Chem Aerosol News 1970; 41:40–43.
4. Bechgaard H, Hagermann NG. Controlled-release multi-units and single unit doses. A literature review. Drug Dev Ind Pharm 1978; 4:53–67.
5. Newton JM. The preparation of spherical granules by extrusion/spheronization. STP Pharma 1990; 6:396–398.
6. Jalal IM, Malinowski HJ, Smith WE. Tablet granulations composed of spherical-shaped particles. J Pharm Sci 1972; 61:1466–1468.
7. Devereus JE, Newton JM, Short MB. The influence of density on the gastrointestinal transit of pellets. J Pharm Pharmacol 1990; 42:500–501.
8. Clarke GM, Newton JM, Short MB. Comparative gastrointestinal transit of pellet systems of varying density. Int J Pharm 1995; 114:1–11.
9. Woodruff CW, Nuessle NO. Effect of processing variables on particles obtained by extrusion–spheronization. J Pharm Sci 1972; 61:787–790.
10. Schwartz JP, Nguyen NH, Schnaare RL. Compaction studies on beads:

compression and consolidation parameters. Drug Dev Ind Pharm 1994; 20:3105–3129.

11. Millili GP, Schwartz JB. The strength of microcrystalline cellulose pellets: the effect of granulating with water/ethanol mixtures. Drug Dev Ind Pharm 1990; 16:1411–1426.

12. Malinowski HJ, Smith WE. Effect of spheronization process variables on selected tablet properties. J Pharm Sci 1974; 63:285–288.

13. Malinowski HJ, Smith WE. Use of factorial design to evaluate granulations prepared by spheronization. J Pharm Sci 1975; 64:1688–1692.

14. Erkoboni DF, Fiore SA, Wheatley TA, Davan T. The effect of various process and formulation variables on the quality of spheres produced by extrusion/spheronization. Poster presentation, AAPS National Meeting, 1991.

15. Chariot M, Francès J, Lewis GA, Mathieu D, Phan Tan Luu R, Stevens HNE. A factorial approach to process variables of extrusion–spheronization of wet powder masses. Drug Dev Ind Pharm 1987; 13:1639–1649.

16. Hileman GA, Goskonda SR, Spalitto AJ, Upadrashta SM. Response surface optimization of high dose pellets by extrusion and spheronization. Int J Pharm 1993; 100:71–79.

17. Goskonda SR, Hileman GA, Upadrashta SM. Development of matrix controlled release beads by Extrusion–Spheronization technology using a statistical screening design. Drug Dev Ind Pharm 1994; 20:279–292.

18. O'Connor RE, Holinez J, Schwartz JB. Spheronization: I. Processing and evaluation of spheres prepared from commercially available excipients. Am J Pharm 1984; 156:80–87.

19. Ojile JE, Macfarlane CB, Selkirk AB. Drug distribution during massing and its effect on dose uniformity in granules. Int J Pharm 1982; 10:99–107.

20. Hellén L, Yliruusi J, Merkku P, Kristoffersson E. Process variables of instant granulator and spheronizer: I. Physical properties of granules, extrudate and pellets. Int J Pharm 1993; 96:197–204.

21. Kleinebudde P, Lindner H. Experiments with a twin-screw extruder using a single-step granulation/extrusion process. Int J Pharm 1993; 94:49–58.

22. Lindberg NO, Tufvesson C, Holm P, Olbjer L. Extrusion of an effervescent granulation with a twin-screw extruder, Baker Perkins MPF 50 D. Influence in intragranular porosity and liquid saturation. Drug Dev Ind Pharm 1988; 14:1791–1798.

23. Harrison PJ, Newton JM, Rowe RC. The characterization of wet powder masses suitable for extrusion/spheronization. J Pharm Pharmacol 1985; 37:686–691.

24. Parker MD, Rowe RC, Upjohn NG. Mixer Torque Rheometry: a method of quantifying the consistency of wet granulations. Pharm Technol Int 1990; 2:50–62.

25. Rowe RC, Parker MD. Mixer Torque Rheometry: an update. Pharm Technol Int 1994; 3:74–82.

26. Baert L, Remon JP, Knight P, Newton JM. A comparison between the extrusion

forces and sphere quality of a gravity feed extruder and a ram extruder. Int J Pharm 1992; 86:187–192.

27. Gamlen MJ. Continuous extrusion using a Baker Perkins MPF 50 D (multi-purpose) extruder. Drug Dev Ind Pharm 1986; 12:1701–1713.

28. Baert L, Fanara D, De Baets P, Remon JP. Instrumentation of a gravity feed extruder and the influence of the composition of binary and tertiary mixtures on the extrusion forces. J Pharm Pharmacol 1991; 43:745–749.

29. Ku CC, Joshi YM, Bergum JS, Jain NB. Bead manufacture by extrusion/ spheronization—a statistical design for process optimization. Drug Dev Ind Pharm 1993; 19:1505–1519.

30. Rowe RC. Spheronization: a novel pill-making process? Pharm Int 1985; 6:119–123.

31. Harrison PJ, Newton JM, Rowe RC. The application of capillary rheometry to the extrusion of wet masses. J Pharm 1987; 35:235–242.

32. Hicks DC, Freese HL. In: Ghebre-Sellassie I, ed. Extrusion and Spheronization Equipment, Chapter 4. Pharmaceutical Pelletization Technology. New York: Marcel Dekker, 1989:71–100.

33. Harrison PJ, Newton JM, Rowe RC. Flow defects in wet powder masses. J Pharm Pharmacol 1984; 37:81–83.

34. Harrison PJ, Newton JM, Rowe RC. Convergent flow analysis in the extrusion of wet powder masses. J Pharm Pharmacol 1984; 37:81–83.

35. Pinto JF, Buckton G, Newton JM. The influence of four selected processing and formulation factors on the production of spheres by extrusion and spheroniza-tion. Int J Pharm 1992; 83:187–196.

36. Hellén L, Ritala M, Yliruusi J, Palmroos P, Kristoffersson E. Process variables of the radial screen extruder: I. Production capacity of the extruder and the properties of the extrudate. J Pharm Tech Int 1992; 4:50–60.

37. Vervaet C, Baert L, Risha PA, Remon JP. The influence of the extrusion screen on pellet quality using an instrumented basket extruder. Int J Pharm 1994; 107:29–39.

38. Baert L, Remon JP, Elbers JAC, Van Bommel EMG. Comparison between a gravity feed extruder and a twin screw extruder. Int J Pharm 1993; 99:7–12.

39. Vervaet C, Remon JP. Influence of impeller design, method of screen performance and perforation geometry on the quality of pellets made by extrusion–spheronisation. Int J Pharm 1996; 133:29–37.

40. Shah R, Kabadi M, Pope DG, Augsberger LL. Physico-mechanical character-ization of the extrusion/spheronization process: Part I. Instrumentation of the extruder. Pharm Res 1994; 11:355–360.

41. O'Connor RE, Schwartz JB. In: Ghebre-Sellassie I, ed. Extrusion and Spheronization Technology, Chapter 9. Pharmaceutical Pelletization Technol-ogy. New York: Marcel Dekker, 1989:187–216.

42. Hasznos L, Langer I, Gyarmathy M. Some factors influencing pellet character-

istics made by an extrusion/spheronization process: Part I. Effects on size characteristics and moisture content decrease of pellets. Drug Dev Ind Pharm 1992; 18:409–437.

43. Hileman GA, Goskonda SR, Spalitto AJ, Upadrashta SM. A factorial approach to high dose product development by an extrusion/spheronization process. Drug Dev Ind Pharm 1993; 19:483–491.

44. Thoma K, Ziegler I. Investigations on the influence of the type of extruder for pelletization by extrusion–spheronization: I. Sphere characteristics. Drug Dev Ind Pharm 1998; 24:413–422.

45. Thoma K, Ziegler I. Investigations on the influence of the type of extruder for pelletization by extrusion–spheronization: II. Extrusion behavior of formulations. Drug Dev Ind Pharm 1998; 24:401–411.

46. Baert L, Remon J. Influence of amount of granulating liquid on the drug release rate from pellets made by extrusion spheronization. Int J Pharm 1993; 95:135–141.

47. Wan LSC, Heng PWS, Liew CV. Spheronization conditions on spheroid shape and size. Int J Pharm 1993; 96:59–65.

48. Newton JM, Chapman SR, Rowe RC. The influence of process variables on the preparation and properties of spherical granules by the process of extrusion and spheronization. Int J Pharm 1995; 120:101–109.

49. Bataille B, Ligarski K, Jacob M, Thomas C, Duru C. Study of the influence of spheronization and drying condition on the physico-mechanical properties of neutral spheroids containing Avicel PH-101 and lactose. Drug Dev Ind Pharm 1993; 19:653–671.

50. Hellén L, Yliruusi J. Process variables of instant granulator and spheronizer: III. Shape and shape distribution of pellets. Int J Pharm 1993; 96:217–223.

51. Dyer AM, Khan KA, Aulton ME. Effect of the drying method on the mechanical and drug release properties of pellets prepared by extrusion. Drug Dev Ind Pharm 1994; 20:3045–3068.

52. O'Connor RE, Schwartz JB. Spheronization: II. Drug release from drug–diluent mixtures. Drug Dev Ind Pharm 1985; 11:1837–1857.

53. Millili GP, Wigent RJ, Schwartz JB. Autohesion in pharmaceutical solids. Drug Dev Ind Pharm 1990; 16:2383–2407.

54. Lustig-Gustafsson C, Kaur Johal H, Podczeck F, Newton JM. The influence of water content and drug solubility on the formulation of pellets by extrusion and spheronisation. Eur J Pharm Sci 1999; 8:147–152.

55. Newton JM, Chow AK, Jeewa KB. The effect of excipient source on spherical granules made by extrusion/spheronization. Pharm Tech Int 1992; 10:52–58.

56. Funck JAB, Schwartz JB, Reilly WJ, Ghali ES. Binder effectiveness for beads with high drug levels. Drug Dev Ind Pharm 1991; 17:1143–1156.

57. Linder H, Kleinebudde P. Use of powdered cellulose for the production of pellets by extrusion/spheronization. J Pharm Pharmacol 1994; 46:2–7.

58. Fielden KE, Newton JM, Rowe RC. The influence of lactose particle size on the

spheronization of extrudate processed by a ram extruder. Int J Pharm 1992; 81:205–224.

59. Mesiha MS, Vallés J. A screening study of lubricants in wet powder masses suitable for extrusion–spheronization. Drug Dev Ind Pharm 1993; 19:943–959.
60. Law MFL, Deast PB. Use of hydrophilic polymers with microcrystalline cellulose to improve extrusion–spheronization. Eur J Pharm Biopharm 1998; 45:57–65.
61. Kleinbudde P. Shrinking and swelling properties of pellets containing microcrystalline cellulose and low substituted hydroxypropylcellulose: I. Shrinking properties. Int J Pharm 1994; 109:209–219.
62. Kleinbudde P. Shrinking and swelling properties of pellets containing microcrystalline cellulose and low substituted hydroxypropylcellulose: II. Swelling properties. Int J Pharm 1994; 109:221–227.
63. Rowe R, Iles C. Spheronization systems downsized for research. Manuf Chem 1995; 11:35–37.
64. Wan LS, Heng PW, Liew CV. Spheronization conditions on shape and size. Int J Pharm 1993; 96:59.

16

Twin-Screw Wet Granulation

Mayur Lodaya and Matthew Mollan
Pfizer, Inc., Ann Arbor, Michigan, USA

Isaac Ghebre-Sellassie
*MEGA Pharmaceuticals, Asmara, Eritrea, and Pharmaceutical
Technology Solutions, Morris Plains, New Jersey, USA*

I. INTRODUCTION

The term granulation, broadly speaking, refers to the process of particle
formation via agglomeration or size enlargement. As described in Perry's
Chemical Engineers' Handbook (1), it is a process of size enlargement where
small particles are clustered into larger, permanent aggregates in which the
original particles can still be identified. The granulation process has been
known and used by mankind since prehistorical times. For instance, mortar
and pestle were used to formulate medicinal agents into solid dosage forms for
the treatment of ailments. A more descriptive definition of granulation (2)
would be "the buildup of clusters from powder or powder/binder mixtures to
produce a free flowing, cohesive material that can be further processed by
compression or encapsulation."

The three general methods of preparing materials for oral solid dosage
forms are direct compression, dry granulation, and wet granulation. The terms
wet granulation and wet massing are often used interchangeably in the
pharmaceutical literature. Granulation is done to improve flow and handling,
bulk density, appearance, solubility, resistance to aggregation, and/or reduction in the propensity for dust formation. Where feasible, direct compression is
the first method of choice for obvious economic reasons. The majority of

actives are not amenable to direct compression due to poor flow or lack of compressibility. In dry granulation, particle buildup is accomplished by compaction or slugging. When applicable, it is often preferred over wet granulation because it involves much fewer processing steps. With the advent of high-speed tabletting machines, an increase in current good manufacturing practice (cGMP) awareness, and an overall drive toward a superior quality of the finished dosage form at a lower cost of manufacture, it has become necessary to develop granulation processes having fewer steps while yielding consistent and high-quality products. Because wet granulation permits the formulation of all types of actives that lack the desired attributes for direct compression as well as those low-dose actives that are difficult to distribute uniformly, it is the most preferred granulation process and is among the most widely used unit operations in the preparation of pharmaceutical solid dosage forms.

Traditionally, pharmaceutical processing is done in batch mode. Some unit operations such as milling, tabletting, encapsulation, etc. are inherently continuous in nature; i.e., as raw materials are fed continuously at a predetermined rate, the product is produced at a specified rate. Others such as wet granulation, drying, etc., as practiced today, are primarily batch operations. In this case, a batch of raw materials is charged to the equipment; all of them are processed simultaneously for a predetermined time under a set of processing conditions to yield the desired product. While continuous processing has long been established in the food, chemical, and plastics industry, this is not so in pharmaceutical processing. The main reasons include tradition, know-how (or lack there of), impact of regulations for any new process, and lack of equipment/ processes that can provide justification in terms of improvement in quality while achieving lower overall cost structure.

Wet granulation as a batch process, along with the advances made in equipment design, has been reviewed periodically (3–5). There have also been a few publications recently that deal with continuous wet granulation (6–11). An attempt will be made in the following sections to review the fundamentals of batch and continuous wet granulation processes, and to discuss the merits of each mode with respect to solid dosage form development and manufacture.

II. DEFINITIONS

A. Wet Massing

Wet massing is an agglomeration technique where powders are mixed with aqueous alcoholic or hydroalcoholic solution and agitated/mixed to produce a pasty product. The final product has a very broad size range starting from less than a millimeter to strands that may be a few millimeters in diameter.

B. Wet Granulation

Wet granulation is the buildup of clusters from powder or powder/binder mixtures to produce, following drying, a free-flowing cohesive material that can be further processed by compression or encapsulation. The particles generally have a low aspect ratio and range from a few hundred microns to less than 1–2 mm in diameter.

C. Continuous Process

Continuous process refers to a process in which there is a continuous feeding of raw materials and a removal of product from the production line.

D. Distributive Mixing

For a given powder blend being processed in a twin-screw extruder (TSE), the term distributive mixing refers to a spatial rearrangement of species that involves the reorientation/generation of new surfaces without any change in the domain size. Domain refers to a unit quantity of material fed to the extruder.

E. Dispersive Mixing

For a given powder blend being processed in a TSE, the term dispersive mixing refers to rearrangement of species through a reduction in the size of domains with/without spatial reorientation.

III. WET GRANULATION IN BATCH MODE

Wet granulation in batch mode is accomplished by shearing a bed of powder blend with mechanical agitation, fluidization or, both. Depending on the amount of shear introduced, the process is described as low shear or high shear. Fluidized bed granulators and some classical mechanical agitation-type granulators fall in the category of low-shear granulators. High-shear mixers composed of higher-agitation blades and chopper constitute the high-shear granulators. Extruders used for wet massing in extrusion spheronization also fall in this category. Currently, high-shear mixers and fluid bed granulators are the most widely used equipment in a batch-type wet granulation processes. In certain instances, high-shear mixers are coupled with vacuum-assisted microwave energy for drying to allow a single-step processing.

A. Wet Granulation Using Fluidized Bed

The fluidized bed granulation of pharmaceuticals was first described by Wurster (3) in 1959. In the following decade, fluidized bed granulators were installed in many pharmaceutical facilities. At present, the equipment is available with capacities ranging from a few grams to several hundred kilograms. It is a very versatile equipment in that mixing, wetting, and drying are all carried out in one equipment. It is one of the most common processing equipment present in pharmaceutical manufacturing floors across the industry. Since their introduction, manufacturers have made significant changes in equipment design to improve the processing efficiency and the robustness of the critical process variables. Some of these advances include the control of fluidizing air in terms of flow and temperature as well as dew point, the nozzle design for granulating liquid flow, the fail safe interlock systems, and the introduction of rotary fluidized bed granulators, among others. Not withstanding all of these improvements, it can be said that high-shear granulators are still the equipment of choice for wet granulation in batch mode, particularly for materials that have low bulk densities. It is practically impossible to achieve the same degree of densification in fluidized bed processes. Fluidized bed granulation processes are also characterized by a long residence time, the need for relatively high amounts of granulating liquid, and the production of high-porosity/low-bulk density granulations. Overall, however, the benefits of fluidized bed granulation far outweigh its disadvantages and the process is still widely used in the pharmaceutical industry. The extensive use of the technology has given users enough experience and empirical data to predictably manage scale-up and end point determination.

B. Wet Granulation Using Low-Shear Granulators

As the name indicates, low-shear granulators are designed to impart much less shear to the powder bed when compared with high-shear granulators. Low-shear granulators are classified into ribbon/paddle blenders, planetary mixers, orbiting screw mixers, and sigma blade mixers. Ribbon blenders and planetary mixers have been more prevalent compared to others. These equipment have been used for dry blending as well as wet granulation. Low-shear granulators produce granules that are denser than those produced in fluid bed. These granulations, however, are fluffier and porous when compared with those obtained from high-shear granulators. In all cases, the wet granulations are either dried in forced air ovens or fluidized bed equipment. In general, low-shear granulation processes lack robustness, and the granulations produced may lack consistency and uniformity at the granule level.

C. Wet Granulation Using High-Shear Mixer Granulator

High-shear mixers, first introduced in the 1970s (4), have been applied to wet granulation, melt granulation, and pelletization. In a high-shear mixer, blending and wet massing are accomplished through high mechanical agitation by an impeller and a chopper. The shear and compaction forces exerted by the impeller provide the energy for mixing, densification, and agglomeration. The chopper, on the other hand, cuts the lumps of wet mass so formed into smaller fragments and aids in the distribution of granulating liquid. High-shear granulation processes are characterized by short residence time and relatively low amounts of granulating liquid, by occasional overwetting due to compaction resulting in uncontrolled granule growth, and by the inability to process physically and chemically unstable materials due to high mechanical shear and high localized temperatures. Scale-up and end point determination are still generally empirical in nature and are highly dependent on the user experience with the type of machine being used. In a recent article, Vromans et al. (12) studied granulation homogeneity as a function of particle size and impeller speed. Results showed that inhomogeneities were found and that they were mainly due to an imbalance of impact pressure exerted by the impeller and the shear resistance of the nuclei. Nevertheless, during the last decade, high-shear granulation processes have proven to be the wet granulation process of choice throughout the pharmaceutical industry. The processing of fragile materials, where there is unusually a high risk of material degradation and where the process operating window is very small, appears to be the only exception.

D. Wet Granulation Using Extrusion Spheronization

Commercialized in the 1960s, the process involves the extrusion of a wetted mass of a mixture of active ingredients and excipients into cylindrical segments that are rounded into spherical pellets in a spheronizer. While pelletization has been studied and reviewed extensively (13), this section will only summarize the extrusion spheronization technique, as wet massing is an integral part of it.

Reynolds (14), Conine and Hadley (15), Jalal et al. (16), and Malinowski and Smith (17,18) did the initial work on pelletization using extrusion spheronization. Pelletization followed by milling was shown to provide several advantages over existing wet granulation techniques employed at the time. The advantages included an improved granulation flow rate, faster drying times, narrow particle size distribution, and lower friability. In spite of these advantages, the use of extrusion spheronization for the manufacture of granulations failed to materialize mainly due to the fact that the process involves several stages (mixing, wetting, extrusion, spheronization, drying, etc.), thereby increasing the cost of products. The process has thus been limited to the production of pellets for controlled-release purposes. In a recent article, Pinto et al. (19) investigated the coextrusion and the spheronization of wet masses. The extrudate comprises concentric cylinders. They claimed that coextruded pellets can provide a means for (a) the simultaneous administration of noncompatible drugs, each present in a different layer; (b) the controlled release of the inner layer; and (c) the modulated release by different amounts of drugs present in different layers.

IV. WET GRANULATION IN SEMICONTINUOUS MODE

Glatt recently introduced the Glatt multicell (GMC) system that falls in the semicontinuous scheme. Leuenberger (20) wrote a review article on the functional aspects of GMC. Figure 1 shows the schematic of such a unit.

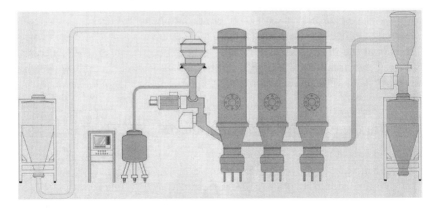

Figure 1   Glatt multicell unit schematic.

The system consists of a metering unit, a high-shear mixer–granulator, a sieve for wet milling, and three fluidized bed dryers. The overall batch is broken in small subunits of 5–9 kg. These subunits are fed successively by the dosing system into the high-shear mixer granulator, which mixes and granulates the materials. The wet granulation is sieved and sequentially conveyed through the series of three fluid bed dryers for drying. While the fluid bed dryers are the same as those used in a conventional batch system, the high-shear granulator has been modified. It has a high-pressure spraying and a dosing system and the tooling design allows a relatively high energy input in comparison to conventional high-shear granulators. The design permits continuous wall cleaning and it is capable of self-discharging. Granulation is done in one stage while drying is accomplished in three sequential steps as the minibatch moves through the system. This system provides the opportunity to optimize the formulation and the granulation process at small scale using the final equipment. It also allows the manufacture of stability batches earlier and avoids the need for multiple biostudies. Because batch quantity is small, the process is inherently designed for better control and robustness. On the other hand, it is a complex system and will require higher initial development efforts to optimize the process for a given formulation.

V. WET GRANULATION IN CONTINUOUS MODE

High-shear granulators, which are the most commonly used equipment for wet granulation batch processing, have some serious shortcomings in that they tend to lead to uncontrolled granule growth and material degradation owing to localized high temperatures. These disadvantages stress the need for a robust wet granulation equipment/process that has a broad applicability and is easy to scale-up. The desired characteristics of one such equipment can be listed as below:

1. applicability for processing powder blends with diverse characteristics;
2. reduction of the level of excipients and granulation liquid required;
3. provision of reproducible processing history resulting in consistent quality from granule to granule;
4. short processing times; and
5. ease of scale-up.

Attempts have been made to modify fluid bed machines and other equipment to achieve these characteristics. In a recent article, Bonde (6) has summarized the technical aspects of continuous processing. He classified the batch and continuous processing equipment as mechanical/fluidized bed/other type, and briefly described continuous fluidized bed granulators. According to him, the lack of continuous processes in the solid dosage form manufacturing is mainly due to tradition, know-how (or lack there of), apparent skepticism toward cGMP compliance as it related to continuous processes, and significant changeover required in the equipment used currently. The paper, however, contends that while batch processing will continue to be important, continuous processing will increase its share. Continuous processes with minimal product holdup and the ability to produce consistently predictable quality product are expected to supplant batch processes in the near future, particularly for high-volume products. This section begins with a short description of the continuous fluidized bed granulators followed by the Iverson mixers and the TSEs. While the continuous fluidized bed and the Iverson mixer have some advantages over the batch processes discussed earlier, they do not meet all of the characteristics described above. The TSE, on the other hand, possesses all of the above attributes and does have the potential to replace high-shear granulators as the preferred equipment for wet granulation. In addition to possessing the desired characteristics mentioned above, the TSE allows process optimization at small scale using the final equipment. This is shown with a discussion on extruder mixing capabilities from examples from early studies and some results from work done in our laboratory. Lindberg (8), Gamlen and Eardly (9), Lindberg et al. (10), and, recently, Keleb et al. (11) have discussed the use of a TSE for continuous granulation. We have successfully used TSE in our laboratory. In some instances, the high-shear and fluid bed processes fail completely while TSE can be used to make consistent granulations. For all continuous processes, operating time becomes a key parameter during scale-up. For example, the equipment used for making clinical batches can, in principle, be operated for longer times to make the stability batches, pivotal supplies, and, finally, the dosage form during manufacturing.

A. Continuous Fluidized Bed Granulators

While continuous fluid bed process technology has been in use for a long time in other industries, none of these systems met the cGMP criteria. Recently, however, a new generation of equipment designed for pharmaceutical processing has been built by Heinen GmbH (Germany), Niro Aeromatic Fielder AG (Switzerland), and Glatt Company (Binzen, Germany). Most of these equipment have five or more functional zones—product in-feed zone, product mixing and preheating zone, spraying zone, drying and cooling zone, and discharge zone. The zones are not necessarily separated mechanically from each other, but the designation is based on the material processing that occurs in a specific portion of the equipment. The material movement occurs via vibration (Heinen), or through the use of air distributor plates that create directed jets (Niro and Glatt).

Heinen Continuous Fluid Bed

Figure 2 shows the Heinen continuous fluid bed processor. It is based on a traditional square bed. The material movement occurs via a vibration of the processing chamber. The product is processed on low expanded fluid bed. The agglomeration liquid is sprayed into the fluidizing powder by means of integrated spray nozzles. A cross stream of warm air, flowing from bottom to top, dries the granulation.

Niro and Glatt Continuous Fluid Bed

Figures 3 and 4 show the illustrations of Niro and Glatt continuous fluid bed granulators. In both these units, the material movement occurs through the use of air distributor plates that create directed jets.

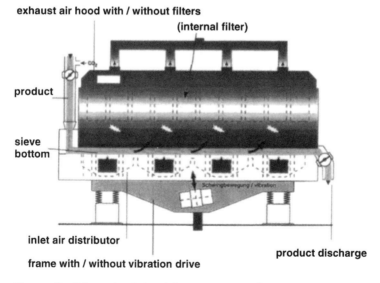

exhaust air hood with / without filters

(internal filter)

product

sieve bottom

inlet air distributor

frame with / without vibration drive

product discharge

Figure 2 Schematic of the Heinen continuous fluid bed processor. (Courtesy of Heinen AG.)

(a)

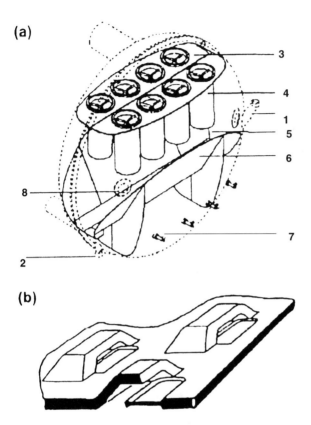

(b)

Figure 3 (a) Reaction chamber of Niro granulator: (1) powder inlet, (2) product outlet, (3) filter outlet, (4) filters, (5) reaction chamber, (6) sieve floor, (7) inlet air chamber, (8) inspection window. (b) Detail of gill plate forming the sieve floor. (From Ref. 7.)

As shown in Figure 3 for the Niro unit, the sieve floor (item 6) divides the granulator into an upper single cell reaction chamber (item 5) and a lower, multicell design, inlet air chamber (item 7). The inlet air chamber is divided into three zones by two vertical walls. The sieve floor has the gill plate construction that helps to produce a directional transport of the contents of the reaction chamber from the inlet (item 1) to the outlet (item 2). Spray nozzles are attached to the sieve floor in bottom spray fashion. The top of the reaction chamber has two parallel rows of sintered metal filters (item 4), four in each row. The inlet air flow and the temperature in each of the three zones of the

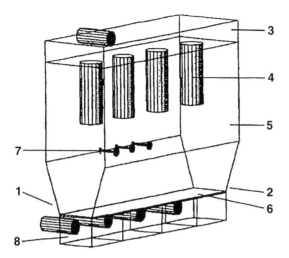

Figure 4 Reaction chamber of Glatt granulator. (1) Powder inlet, (2) product outlet, (3) filter outlet, (4) filters, (5) reaction chamber, (6) sieve floor, (7) spray nozzle, (8) inlet air chamber. (From Ref. 7.)

inlet air chamber are separately specified to provide spatially varying conditions for mixing, agglomerating, and drying in the overlying reaction chamber. The powder enters into the first region of the reaction chamber where it is mixed. Passing onward, the granulation liquid is sprayed onto the powder and agglomeration occurs. In the last zone, the granules are dried and transported out of the reaction chamber.

As shown in Figure 4, the Glatt unit also has two horizontal chambers. These are the inlet air chamber (item 8) and the reaction chamber (item 5). Three vertical walls in the inlet air chamber divide it into four zones, in each of which the temperature and the air flow are separately controlled. The major difference with the Niro unit is the absence of the gill plate. Instead, the sieve floor (item 6) has uniformly placed round perforations and is not designed to provide a directional transport of the powder, similar to the Niro unit. Four metal filers (item 4) are located on top of the reaction chamber. The powder fed at the inlet (item 1) is mixed in the first zone, agglomerated by top spray in the second zone as well as third zone, and dried in the fourth zone before it exits the reaction chamber from the outlet (item 2).

Silke et al. (7) examined the movement of powder granules in the Niro and Glatt units to determine if these machines provided directional transport or random mixing within the processing chamber. It was concluded that while the targeted air flow does result in material movement from the inlet to the exit, it does not prevent random mixing. In essence, it is quite possible that the processing history of the individual granules would be quite different from each other, depending on the respective residence times. The average residence time reported was on the order of 20–25 min, which is high. Altering process conditions, such as spray rate or air flows, produced predictable changes in granule properties.

B. Iverson Mixer

Lindberg (8) used an Iverson mixer to study wet granulation of several placebos as well as active formulations, and concluded that the mixer is a suitable alternative to planetary and high-shear mixers for granulation. Powders and liquid are metered into a narrow space at the periphery of the grooved disc, which rotates at high speed (Figure 5). Powder is metered onto the upper part of the disc. A single powder feeder (containing a preblend) or multiple powder feeders with individual components (all feeding onto the disc, can be used. The liquid is pumped to the lower side of the rotor. Disc rotation transports the liquid to the periphery of the disc, and drops are formed at the rim. The granulation takes place in the narrow space where the powder stream engages into the liquid droplets. The overall residence time is very short (less than 1 sec). The modified Iverson mixer used in these experiments corresponded to a NICA M50.

The granulations of an active as well as a placebo (lactose, starch, and polyvinyl pyrrolidone), with powder metering rates of 300 and 600 kg/hr, were prepared. Varying amounts of time, from 5 to 30 min were explored. The resulting granulation was dried in batch mode (using a batch fluidized bed or tray dryer) and continuous mode (using vibro fluidized bed dryer). The granulations had the desired properties and were consistent when the ratio of powder/liquid flow was kept same.

C. Twin-Screw Extruders

As has been discussed in Sec. VB, TSEs offer several advantages over other wet granulation processes. The modular nature of the screw elements and the large variety available provide the user with tremendous flexibility. Varying the levels of distributive and dispersive mixing allows the generation of granulations with the desired characteristics by controlling the degree of agglomeration, particle buildup, and homogenization. The screws interact with

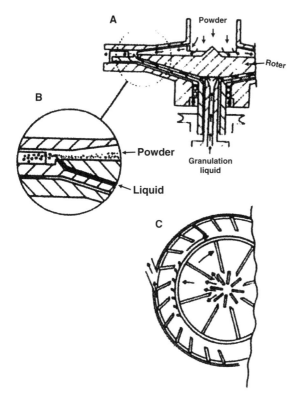

Figure 5 Schematic of the Iverson mixer. (A) Side view showing, rotor, liquid, and powder inlets. (B) Magnified view of the agglomeration space. (From Ref. 8.)

each other to provide self-cleaning, enhanced pumping, and different levels of mixing. Dispersive mixers break down the morphological domains, whereas the distributive mixers space the morphological domains without altering them. A small mass processed at any given time, along with short local mass transfer distances, results in a very controlled and reproducible shear and processing history from granule to granule. The average residence time is very short (on the order of 1 min) when compared to other granulation equipment. The modular nature of the barrels provides the needed flexibility in accomplishing several unit operations within the process section. The process section is divided into feeding zone, mixing zone, and discharge zone, as discussed by Lindberg (8). All of the above discussed characteristics combine to give a broad applicability of TSE for the manufacture of wet granulation. The product is either extruded in the form of a spaghetti, or as discrete granules. Other key operational benefits include a reduction in the level of excipient, an overall flexibility in the output, and a simple visual end point determination. As will be described in the general process description, a quick adjustment of solid and liquid feed rates during the start-up phase is required to achieve the desired quality of granulation.

Lindberg (8), Gamlen and Eardly (9), and Lindberg et al. (10) have discussed the use of a Baker Perkins MP50 mixer/extruder for continuous granulation. More recently, Keleb et al. (11) used a 25-mm APV Baker TSE to produce granulation in the continuous mode. Figure 6 shows a typical feeding/mixing and discharge zone as described by Lindberg (8). Paddles (kneading discs) were inserted in the mixing zone to provide intense shear mixing. The screw elements used in the feed zone convey the product, while the screw elements used in the discharge zone develop the die pressure while conveying the product. The granulation was extruded in the form of spaghetti ranging in size from 0.5 to 1.0 cm in length. The key process parameters were screw configuration, liquid and powder flow rate, screw speed, and die pressure. In a separate publication, Lindberg et al. (10) reported the granulation of an effervescent formulation (containing anhydrous citric acid and sodium bicarbonate) using ethanol as the granulating liquid in the same equipment.

Gamlen and Eardly (9) prepared a granulation of paracetamol with high drug loading. They evaluated the influence of formulation composition and moisture content on the quality of extrudate obtained from the Baker Perkins MP50 extruder. In particular, they examined the effect of adding a small

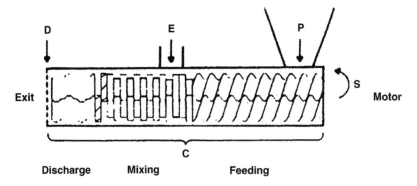

Figure 6   Schematic of TSE showing the processing zones. (From Ref. 8.)

amount of hydroxy propyl methyl cellulose to a granulation of paracetamol, Avicel, and lactose. They observed that the addition of hydroxy propyl methyl cellulose improved the extrudability. A moisture range of 20–25% provided acceptable extrudates. Their overall conclusion suggested that additional formulation work was needed to overcome the surface roughness of the extrudate.

Keleb et al. (11) examined the use of TSE as a single-step granulation/ tabletting device. They produced 9-mm-diameter extrudates, cut them manually into 4-mm-thick tablets, and dried them for 20 hr at 25°C. The formulation used contained hydrochlorothiazide as a model drug, with lactose monohydrate as a diluent and polyvinylpyrrolidone (PVP) and water as binder. They concluded that single granulation/tabletting step using extrusion could be used as an alternative tabletting technique for actives with poor compression properties.

General Process Description and Process Parameters

As with any twin-screw application, depending on the level of mixing required, the first step is to determine a screw design appropriate for granulation. Knowing the water-carrying capacity of ingredients present in the formulation and the desired moisture level, the overall ratio of powder to liquid feed can be decided. At the start of the process, powder feed is started at low levels, whereas the liquid feed is started at comparatively higher levels. The screw speed is set at low revolutions per minute (rpm). As would be expected, the initial output is an overwetted material. The screw speed, the powder speed, and the liquid feed are all ramped up in small but quick steps until the desired throughput of proper granulation is observed at the outlet. There is a working window of at least a few percentage points for the liquid feed in which acceptable granulation is achieved. The most important response variables during startup are motor torque and the quality of granulation, as observed visually. Because the residence time is very short, in general, it is possible to come to steady state operation in a few minutes from the start. Steady state is measured in terms of a steady torque, stable temperatures, and an acceptable granule quality.

Independent Process Variables. Key independent process variables are screw configuration, screw rpm, temperature, and feed locations for liquid feed. The screw design influences the granulation characteristics and the overall processability for a given formulation. Any screw design selected for granulation will have a feeding zone, one or more mixing zones, and a discharge zone. A few discrete mixing zones are used to achieve the desired level of mixing, agglomeration, and homogenization. The conveying elements in the feed zone are designed to capture the material fed to the process. For a 25–34 mm TSE, the achievable dry powder blend throughputs range from 4–6 to 25–30 kg/hr, depending on the blend characteristics and screw design employed. The screw range of 100–300 rpm, normally used for 27–34 mm machines, was used in our studies and was found to be adequate. However, in the newer extruders being marketed, it is common to operate up to 500 rpm.

Dependent Process Variables.   Key dependent process variables are powder feed rate, liquid feed rate, and motor torque. As for the motor torque, the key is to operate at a steady torque. For a given screw design and screw rpm, the maximum powder feed rate is defined by the rate at which the torque is 80% of the manufacturer-recommended limiting torque. Depending on the moisture-carrying capacity of the powder blend, the maximum liquid feed rate is defined. As long as the powder feed and the liquid feed are steady, it is fine to operate at about 80% of the manufacturer-recommended limiting torque.

For powder feed, a side feeder is used. Alternatively, the powder blend can be gravity-fed. For liquid feed, depending on the characteristics of the powder feed in question, it may be advantageous to divide the overall amount into two or three locations.

TSE Granulation Examples

Several granulations consisting of commonly used excipients such as lactose, microcrystalline cellulose (MCC), dicalcium phosphate, etc. were prepared using a TSE. Table 1 shows the composition of some of these formulations. These formulations were successfully granulated at moisture levels varying from 20% to 30% to as high as 40–50%. The MCC-based formulations took up higher moisture levels due to its high water-carrying capacity. Some of

Table 1   Sample Formulation Composition for Placebo Granulations

Ingredients	Formulation composition (%)		
	Formulation A	Formulation B	Formulation C
MCC	66	32	
Lactose	32	66	
Dicalcium phosphate			98
Cross povidone	2	2	2

these feasibility studies were run continuously for over 8 hr, at powder blend feed rates of 20–30 kg/hr using a 34-mm twin screw mixer (TSM).

In addition, feasibility studies involving formulations of several investigational compounds have been conducted. In some of these, the studies demonstrated that successful granulation could only be done using TSE, whereas high-shear and fluid bed systems failed to provide a consistent quality product.

Investigational Compound A. Several different granulation methods were attempted. These included high shear, roller compaction, high shear followed by roller compaction, and melt extrusion. None of these methods was successful in providing a granulation that could be compressed into a dosage form, yielding a consistent release of the drug as required. In the case of high-shear granulation, the final product was taffylike consistency and very difficult in terms of further processing. Using TSE, on the other hand, consistent granulations were prepared with the same formulation and with a higher level of actives.

Investigational Compound B. In this case, the high-shear and fluid bed processes were initially evaluated for wet granulation. The high-shear process was abandoned as it was very difficult to wet the material. Due to very low bulk density, the active was found to ride up in the bowl. The fluid bed process seemed to work fine on small scale. However, during scale-up,

Figure 7 Investigational compound B: fluid bed granulation—SEM of granular cross section.

Figure 8 Investigational compound B: TSM granulation—SEM of granular cross section.

the granulation exhibited poor flow and picking was observed during compression. Wet granulation using TSE was developed as an alternate. Figures 7 and 8 show the scanning electron microscopy (SEM) of the two granulations. For the fluid bed granulation, there is a clear evidence of weakly granulated and ungranulated active pharmaceutical ingredient (API) crystals. For TSE granulation, on the other hand, the SEM shows a complete coverage of active with the excipients. Compression studies with TSE granulation proceeded satisfactorily. For the tablets made, Table 2 and Figure 9 show physical testing results and compression profiles, respectively.

Table 2 Investigational Compound B Tablets: Physical Testing Results

Compression force [kg]	Hardness [kP] (n=10)	Thickness [mm] (n=10)	Weight variation [mg] (n=10)	Disintegration time [min] (n=3)	Friability [%] (n=20; time=10 min)
526.3	9.1±0.5	0.247	644.1±4.1	2.0	0.88
731.3	14.2±0.8	0.237	651.2±4.7	3.0	0.35
823.1	16.3±0.9	0.232	649.4±4.1	3.7	0.32
945.5	19.3±1.3	0.227	649.2±4.2	4.7	0.18
1200.5	24.2±1.4	0.22	649.2±5.4	5.3	0.18

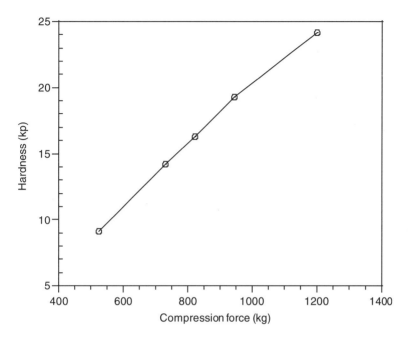

Figure 9 Compression profile for investigational compound B.

In the forgoing discussion, an attempt is made to demonstrate the utility of TSE in the preparation of wet granulation. Obviously, many of the fundamental research works need to be done to compare it with conventional equipment such as high-shear granulators and fluid bed.

VI. CONCLUSIONS

Continuous wet granulation processes have definite advantages over the batch processes. These include improved quality, improved manufacturing space utilization, reduced development time, and overall reduced cost. Improved quality is achieved by incorporating in-process controls and real-time monitoring of the process. By reducing the number of process steps and staging as well as overall handling of the in-process materials involved, manufacturing space is utilized more efficiently. Because fewer scale-up steps are involved, the overall time scale of development is reduced. The same equipment used for making Phase III clinical supplies can be used for manufacturing scale by running it for a longer time. All of the above, combined together, provide for a reduction in development and manufacturing cost. However, not only does the continuous process need to be inherently robust, it requires an accurate feeding system (for both the powder and the liquid) to maintain a uniform quality of the product. Also, historically, continuous processes have been proven advantageous for high-volume products in other industries. In pharmaceutical processing, at least initially, the continuous processes will gain acceptance only on the basis of a significant quality advantage.

REFERENCES

1. Green DW, Maloney JO, eds. Perry's Chemical Engineers' Handbook. New York, NY: McGraw-Hill Publication, 1997.
2. Kadam KL, ed. Granulation Technology for Bioproducts. Boca Raton, FL: CRC Press, Inc., 1990.
3. Kristensen H, Schaefer T. Granulation—a review on pharmaceutical wet granulation. Drug Dev Ind Pharm 1987; 13(4&5):803–872.
4. Holm P. High shear mixer granulators. In: Parikh DM, ed. Handbook of Pharmaceutical Granulation Technology. New York, NY: Marcel Dekker, 1998:151–203.
5. Parikh D, ed. Handbook of Pharmaceutical Granulation Technology. New York, NY: Marcel Dekker, 1998.
6. Bonde M. Continuous granulation. In: Parikh DM, ed. Handbook of Pharmaceutical Granulation Technology. New York, NY: Marcel Dekker, 1998:369–386.
7. Silke G, Knoch A, Lee G. Continuous wet granulation using fluidized-bed techniques: I. Examination of powder mixing kinetics and preliminary granulation experiments. Eur J Pharm Biopharm 1999; 48:189–197.
8. Lindberg NO. Some experience of continuous granulation. Acta Pharm Suec 1988; 25:239–246.
9. Gamlen MJ, Eardly C. Continuous extrusion using a Baker Perkins MP50 extruder. Drug Dev Ind Pharm 1986; 12(11–13):1701–1713.
10. Lindberg NO, Myrenas M, Tufvesson C, Olbjer L. Extrusion of an effervescent granulation with a twin screw extruder, Baker Perkins MPF 50D determination meant time. Drug Dev Ind Pharm 1988; 14(5):649–655.
11. Keleb EI, Vermeire A, Veravaet C, Remon JP. Cold extrusion as a continuous single-step granulation and tabletting process. Eur J Pharm Biopharm 2001; 52:359–368.
12. Vromans H, Poels-Janssen HGM, Egermann H. Effects of high shear granulation on granulate homogeneity. Pharm Dev Technol 1999; 4(3):297–303.

13. Ghebre-Sellassie I. Handbook of Pelletization Technology. NY: Marcel Dekker, Inc., 1989.
14. Reynolds AD. Manuf Chem Aerosol News 1970; 41:40.
15. Conine JW, Hadley HR. Drug Cosmet Ind 1970; 106:38.
16. Jalal IM, Malinowski HJ, Smith WE. Tablet granulations composed of spherical shaped particles. J Pharm Sci 1972; 61:1466–1468.
17. Malinowski HJ, Smith WE. Effect of spheronization process variables on selected tablet properties. J Pharm Sci 1974; 63(2):285–288.
18. Malinowski HJ, Smith WE. Use of Factorial design to evaluate granulations prepared by spheronization. J Pharm Sci 1975; 64(10):1688–1692.
19. Pinto JF, Lameiro MH, Martins P. Investigation on the co-extrudability and spheronization properties of wet masses. Int J Pharm 2001; 227:71–80.
20. Leuenberger H. New trends in the production of pharmaceutical granules: batch versus continuous processing. Eur J Pharm Biopharm 2001; 52:289–296.

17

Installation, Commissioning, and Qualification

Adam Dreiblatt

Extrusioneering International, Inc., Randolph, New Jersey, USA

I. INTRODUCTION

It is not within the scope of this chapter to describe validation concepts in general, but rather, to identify those aspects of melt extrusion that differentiate it from the validation of conventional pharmaceutical equipment and processes. Because melt extrusion technology has not yet gained widespread acceptance as a traditional pharmaceutical unit operation, the installation, commissioning, and validation of extrusion equipment have not been widely discussed in the pharmaceutical literature (1). Extruders are rather unfamiliar to those writing validation protocols, creating problems in both documentation and expectations for the users of such equipment. It is the objective of this chapter to provide an overview for developers of validation protocols as well as for project engineers responsible for installing and commissioning melt extrusion equipment.

Extruder manufacturers have only been supplying equipment to the pharmaceutical industry for a relatively short period of time. The equipment was originally designed for use in the plastics processing industry and has been modified to suit the needs of the pharmaceutical industry. While suppliers of "traditional" pharmaceutical equipment are intimately familiar with cGMP requirements in terms of machine design and documentation, extruder manufacturers are still on the learning curve as far as identifying the unique needs of this industry. It is a precarious situation whereby the equipment suppliers are not familiar enough with cGMP requirements and the users are not familiar enough with the equipment. It is this difference which makes the commissioning and the qualification of extrusion equipment more complicated than normal.

II. INSTALLATION

Some aspects of the melt extrusion process, which may not be obvious but require special attention for a successful installation, are listed below.

A. Floor Space Requirements

One major difference between extruders used for pharmaceutical applications and those used for "conventional" plastics processing is in the solubility of the polymers used in the formulations. Most thermoplastics are hydrocarbon-based and are relatively insoluble as a result. This property allows water to be used in direct contact with the extruded material as a cooling medium to solidify the melt for further processing. Plastics are extruded as molten strands and are typically drawn through a water trough to cool the material below its T_g for pelletizing. Direct contact with water provides efficient cooling, thereby minimizing the residence time required in water.

Because most of the polymers used in pharmaceutical applications are water-soluble or have some degree of water solubility, direct contact with water for cooling and solidification of the extruded melt is not possible. This implies cooling in air or using indirect water-cooled surfaces (e.g., cooling belts and/or chilled rolls) and longer residence times for cooling. The consequence is increased floor space requirements for these applications. It is not uncommon for downstream cooling equipment to take up more than three to four times as much floor space as the extrusion equipment. This does not include the subsequent pelletizing equipment, which would follow the cooling step. The extruded strands must be in-line with the extruder, or the die can be oriented 90° to extrude at a right angle from the extruder centerline. In either case, a substantial amount of length may be needed downstream of the extruder. Thus the floor space for melt extrusion applications should be considered prior to installation in an existing area.

B. Ceiling Height

While the extruder discharge is in a horizontal plane (the standard centerline height for twin-screw extruders is approximately 1.1 m), the extruder itself does not require very much floor space nor does it require much headroom. It is the continuous operation of the feeding equipment situated directly over the extruder that can impose some height restrictions. Because the extruder is a continuous device, the feeder will need to be refilled several times. Refilling the feeder can be done manually or automatically; it is the automatic refill systems which can require considerable headroom above the feeder (refill hopper plus associated conveying equipment and refill valves).

C. Fugitive Dust

The material discharged from the feeder(s) into the extruder may produce some fugitive dust at the feed opening of the extruder. The refilling of the feeder may also produce some fugitive dust (especially for manual refill systems). The degree to which this dust becomes airborne depends on the material. Standard dust extraction hoods and systems should be considered in the vicinity of the feeder inlet port as well as the extruder feed opening. The extruder feed opening must be permitted to "breathe" to remove the air introduced into the extruder via the powder feed.

D. Fume Exhaust

Depending upon the materials being processed, there may be some fumes present at the extruder discharge as a result of the material being exposed to the room environment at an elevated temperature. Whether fume extraction hoods are required will depend on the nature of the materials being processed and the product temperature at the discharge.

E. Auxiliary Equipment

Auxiliary support equipment for the extruder (water circulator for barrel cooling, vacuum pump, etc.) can be located outside of the process area for convenience. Utility connections can be made at wall-mounted panels to connect to the extruder via quick-connect fittings and flexible hoses.

III. COMMISSIONING

The act of commissioning for process equipment is neither well defined nor regulated; recent attempts have been made to standardize the commissioning methodology, the terminology, and the documentation practices (2–4). Unless the time, resources, and responsibilities are specified in the Validation Master Plan (see Sec. IV), the commissioning process becomes a "make it work" effort with no clearly defined objectives or schedule. It is often left up to the vendor to debug the system after installation and before qualification. Sufficient resources and time must be allocated in the project schedule to coordinate commissioning efforts with the construction/installation and validation activities. It is also important to remember that all equipment must be commissioned; however, only equipment which has a direct impact upon, or is critical to product quality needs to be validated.

In nonpharmaceutical applications of extrusion equipment, the commissioning process is called "start-up." For the commissioning of melt extrusion equipment for pharmaceutical applications, commissioning includes start-up, operator training, instrument calibration, verification of "as-built" documentation, etc. The commissioning of melt extrusion equipment provides an opportunity for operators, engineers, and maintenance personnel to become familiar with the equipment. At this point, there are still minor modifications that may be made to the equipment and/or installation prior to executing IQ protocols. Commissioning is also an opportunity to revise Installation Qualification/Operational Qualification/Performance Qualification (IQ/OQ/PQ) protocols based upon the "as-built" installation and/or equipment.

IV. QUALIFICATION

The equipment involved in the melt extrusion process is still required to comply with cGMP regulations covered in CFR 21, Part 211, "...equipment used in the manufacture, processing, packing, or holding of a drug product shall be of appropriate design, adequate size, and suitably located to facilitate operations for its intended use and for its cleaning and maintenance (5,6)."

The validation process is no different for an extruder; the Validation Master Plan sets the scope and the strategy for validation activities (7):

Introduction (objective, validation scope, etc.)
Definition of validation concepts (identify systems which require validation, definition of IQ, OQ, PQ, etc.)
Process description (design criteria, flow diagrams, material balance, process instrumentation and control, etc.)
Facility description (overall layout, area classification, material/personnel flow, HVAC requirements, etc.)

Validation activities [process systems, utility systems, calibration, standard operating procedures (SOPs)]

Validation approach (acceptance criteria, assumptions, approval procedures, etc.)

Validation schedule

Validation supplies

Validation organization.

An attempt will be made to identify those "issues" relating specifically to the unique nature of melt extrusion as it pertains to qualification (IQ/OQ/PQ).

It has already been mentioned that melt extrusion is a relatively new process for the pharmaceutical industry. As such, there is a lack of *fundamental* understanding as far as both process and the equipment are concerned (note that this fundamental understanding has been developed within the plastics processing industry over several decades). Without a good understanding of the melt extrusion process, a User Requirements Specification (URS) is difficult to create.

Without a URS, the equipment supplier's quotation then creates the specifications against which protocols are developed. This situation must change, as the validation strategy is flawed if the equipment supplier is determining the specifications for the process. At a minimum, a Specification Qualification (SQ) and/or Design Qualification (DQ) step must be included within the Validation Master Plan, where the melt extrusion process requirements are matched to equipment specifications. The vendor's equipment proposal must meet the specifications required for the process, rather than determine the specifications for the process (8).

The Validation Master Plan must also account for Factory Acceptance Testing (FAT), an often overlooked yet important step in the project life cycle (9). Without experienced equipment specialists on-site, it becomes more difficult to resolve any technical problems after the equipment has been delivered and installed. The FAT provides an opportunity to identify any faults, prequalify draft SOPs, as well as verify conformance to SQ and/or DQ. A FAT protocol with extensive testing can be used to minimize the IQ/OQ effort on-site. The FAT should include dry and/or wet testing. An increased pressure to meet project deadlines is usually responsible for compromising the scope of the FAT.

A. Installation Qualification

The IQ protocol(s) is developed to challenge the extruder design and the installation, and is typically conducted in a nonoperating mode (i.e., with no power applied). There are several minor differences between an extruder used for melt extrusion and traditional pharmaceutical processing equipment which need to be addressed in an extruder IQ. Some of the typical IQ requirements are listed below.

Equipment Identification

No differences are noted; extruder nameplate data (manufacturer, model number, serial number, year of construction, etc.) are typically installed on the gearbox.

Lubricants

"... Any substance required for operation, such as lubricants or coolants, shall not come into contact with components, drug product containers, closures, in-process materials, or drug products so as to alter the safety, identity, strength, quality, or purity of the drug product beyond the official or other established requirements (5)."

The extruder gearbox requires gear oil, which is contained within the gearbox housing and is isolated from product contact. Food-grade lubricants, which meet the requirements of the gearbox manufacturer, are available. No other lubricants are required.

Product Contact Materials of Construction

"... Equipment shall be constructed so that surfaces that contact components, in-process materials, or drug products shall not be reactive, additive, or absorptive so as to alter the safety, identity, strength, quality, or purity of the drug product, beyond the official or other established requirements (5)."

This is one area in which extruders differ from traditional pharmaceutical process equipment. Due to the high mechanical stresses involved whereby screws rotate at a relatively high speed (nominal 300 rpm) with close tolerances (nominal 0.2 mm), it is not possible to manufacture extruder screws and barrels out of 304-type or 316-type stainless steels (refer to Chapter 2 for further information on extruder materials of construction). These components are typically fabricated from hardenable grades of stainless alloys, similar to those used for tablet tooling (e.g., type 440B stainless steel) and are subject to corrosion if not handled appropriately. Other product contact components that do not experience high mechanical stress (feed hoppers, vent hardware, etc.) can be fabricated from 304-type or 316-type stainless steels.

Other possible product contact surfaces are screw shaft seals, typically fabricated from Teflon-type materials. Documentation supporting product contact materials of construction is available from the extruder suppliers.

Utility Requirements

Most extruder drives require three-phase power, and are available for either 208, 230/240 or 460/480 Volt service. Most drives are provided with forced air cooling blowers, which are exhausted into the processing room. Cooling blowers and ductwork can be installed to bring air from outside the processing room and exhaust outside of the processing room.

Compressed air is rarely needed, and may be used only on very large extruders for torque-limiting couplings.

The extruder barrels are typically heated using electrical heaters and cooled using a circulating coolant. The circulating coolant does not come in contact with the product and is typically water from a cooling tower or from a dedicated water circulator. The specifications for water quality are available from the extruder supplier. As the water is in direct contact with the extruder barrel, it is possible to develop some corrosion products over time. The quality of this cooling water can adversely affect the process, if sediments that can block the solenoid valves used to control the flow of water through the extruder barrels are present. The installation of a filter in the circulating water system is recommended to avoid this situation. The cooling water should have a temperature control loop; cooling water temperature can vary from $10\,^{\circ}\text{C}$ to $90\,^{\circ}\text{C}$.

If a vacuum pump is used as part of the extrusion system, this will also require water to maintain a vacuum seal, as well as a drain. In some instances, effluents from the extruder vent (e.g., noncondensable gases) will also be discharged to the drain, potentially creating some environmental issues.

The support equipment (water circulator, vacuum pump) can be located outside of the processing room, providing the required utilities through a wall-mounted panel.

Calibration of Critical Instruments

The critical instruments for a melt extrusion process must be identified. Instruments directly associated with the determination of product quality, which are recorded in a Manufacturing Batch Record, are considered critical; all other instruments are regarded as noncritical. Noncritical instruments should also be calibrated for a new installation.

Those instruments that indicate and/or directly control the extrusion process include controlled parameters and response parameters:

1. *Controlled parameters* are parameters which have a set point and/ or a range of set points, including:
   Screw speed
   Extruder barrel/die temperatures
   Vacuum level
   Feed rate.
2. *Response parameters* are parameters whose value depends on the controlled parameters:
   Die pressure
   Product temperature
   Extruder load (torque).

There are no unique features for instruments on an extruder that would present any difference from conventional pharmaceutical equipment in terms of calibration.

Preventive Maintenance

There is nothing unique about the preventive maintenance of an extruder. The documentation accompanying extruders includes sufficient preventive maintenance information. As extruders are typically operated in a continuous environment, the preventive maintenance intervals may appear to be rather large as compared to traditional pharmaceutical process equipment.

Standard Operating Procedures

The procedures required for setup, operation, and cleaning of extruders require some special attention. The cleaning of an extruder used for melt extrusion poses the following challenges, which are typically not encountered with traditional pharmaceutical processes:

1. The extruder must be disassembled for cleaning while heated to a temperature above the T_g of the polymer. This can be up to 200°C for some polymers, providing unique safety hazards for the operating staff.
2. The components must be cleaned of molten polymer, versus powders typically encountered in traditional cleaning processes. The polymers must be dissolved, degraded, or burned to remove the residues. For controlled-release applications utilizing polymers

having low solubility, there are additional challenges to find a suitable cleaning agent if the molten material is not water-soluble.

3. The disassembling of the entire extruder process section (including barrels, screws, etc.) is appropriate when changing to a different product or strength to avoid any possible cross contamination. This can be a very time-consuming process requiring some mechanical aptitude. Because the screws in most twin-screw extruders is withdrawn from the discharge end of the barrel, the internal barrel surface cannot be visually inspected for cleanliness. Clamshell or split-barrel extruders, which may reduce some of the complexity, are available. The assembly of the segmented screws requires an equal amount of time and mechanical aptitude. Partial cleaning (e.g., between lots) can be accomplished without a disassembly of the screws or barrels.

4. As already mentioned, the major extruder components (barrels and screws) are fabricated from hardenable stainless alloys. These components will require "special" handling for cleaning, similar to tablet tooling; they will show signs of corrosion if left exposed to moisture or humidity. These components must be dried thoroughly after cleaning and maintained in a controlled humidity environment.

Because most extruders employ electrically heated barrels, standard extrusion equipment is not normally designed for washdown. If required, extruders can be designed where the barrel temperatures are controlled using hot water or other heat transfer fluids in order to provide a washdown environment.

Safety

Other than the previously mentioned issues of handling the extruder components while in a heated state, there are no other major safety issues other than site-specific requirements. As extrusion equipment is designed for an industrial environment, machine guarding for pinch points and hot surfaces and lockout features must meet appropriate regulatory standards (OSHA, UL, CE,* etc.). Noise levels are typically within industry standards (less than 85 dBA).

---

* OSHA=Occupational Safety & Health Administration; UL=Underwriters Laboratories; CE= mark certifying conformity to European Directives.

B. Operational Qualification

The objective of OQ is to challenge the operational controls, alarms, and interlocks for a given piece of equipment. The systems that should be challenged for melt extrusion need to cover the range of operation for the intended application. Thus if the processing temperature for a given polymer/active formulation is 100°C, then the extruder temperature controls must be challenged beyond this temperature (e.g., up to 150°C) in order to accommodate the operation within the limits of OQ.

Motor Rotation

Because most extruder drives are three-phase, the correct rotation must be verified. Correct motor rotation must be confirmed for all other three-phase motors including main drive motor cooling blower, vacuum pump, control panel cooling fans, etc.

Operating Controls

The controlled parameters described for calibration must be challenged for both functionality as well as for the range of operation. The controlled parameters and associated controls include main power disconnect, main drive motor stop/start functionality, range of operation of screw speed (with specified tolerance), vacuum pump stop/start, and range of operation of vacuum level (with specified tolerance). As previously mentioned, the testing range for temperature control of the extruder barrels and die should be outside of the intended operating range for the melt extrusion process. Both heating and cooling functions are to be challenged, with some acceptance criteria for accuracy.

Because twin-screw extruders are modular in design, with an independent control of each barrel module, the temperature limits for each temperature control zone can be different from each other. This represents an additional degree of freedom for challenging the temperature control of the extruder. If the set points for all temperature control zones are tested at a constant value, the functionality (i.e., ability to control and to maintain set point within some degree of accuracy) of the heating and cooling systems can be verified. When the heating controls are to be tested, the extruder feed barrel can be exempted, as this barrel module is typically controlled with manual cooling only and does not have heating capability. When the cooling controls are to be tested, the dies (and associated die adapters) are typically heated only and can be exempted, as these do not have cooling capability.

The best method to test the heating and cooling systems is to verify the accuracy of each temperature control zone to maintain a set point within some acceptance criteria (e.g., $\pm 5\,^\circ$C) over some specified time interval. The data can be recorded in tabular format for each temperature control zone, starting with the time when the set point is entered. The actual temperature is recorded every 10 or 15 min over several hours to verify not only accuracy but also temperature stability. The temperature control zones can be tested individually; however, in actual practice, the extruder barrels will all be either heated or cooled to a specific temperature profile. All temperature control zones can be set to the same set point, with several set points being tested both above and below the intended operating range for the melt extrusion process.

Alarms/Interlocks

The standard tests for emergency stop switches must be included in this section of the OQ. All extruders have specific alarm and interlock circuitry to prevent damage to the gearbox, which must be challenged:

1. Melt pressure is set up as an alarm to warn of high melt pressure at the extruder die. These are electronic signals from a pressure transducer that can be simulated to verify the alarm circuitry.
2. Melt pressure is also set up as an interlock to shut down the extruder drive when melt pressure exceeds the manufacturer's recommended maximum limit. A signal can be generated to activate and to verify the interlock circuitry.
3. Motor load (torque) limits are set up either as an alarm and/or interlock to shut down the extruder drive. This is one condition that may not be able to be simulated, as the signal for motor load comes directly from the drive. If this is the case, then the extruder must be run with a material and a high motor load condition must be created to verify the interlock circuitry. This test is potentially damaging to the extruder if the alarm/interlock was not set up correctly by the manufacturer. It is strongly recommended that this test be conducted during FAT with the vendor present, or to have the vendor technical specialist witness this test during OQ. One method to impose high motor load is to operate the extruder with polymer and to slowly decrease the extruder barrel temperatures to increase the melt viscosity. Using this method, the motor load limits can be approached gradually.
4. Torque-limiting couplings may also be installed to mechanically isolate the drive motor from the extruder gearbox in the event of an

overload situation. The functionality of these devices also needs to be verified. Because these are mechanical devices, the only way to test their functionality is to actually put them under load (see above procedure to verify motor load interlock).

5. Gearbox oil temperature and pressure may be set up to either signal an alarm condition, or as an interlock to shut down the drive motor in the event of low or high oil pressure and high oil temperature. Signals can be generated to simulate these conditions and to verify the alarm/interlock circuitry.

6. The feeding equipment for the extruder, while not directly part of the extruder itself, must be tested for accuracy and precision. The test procedure for determining accuracy and precision involves operating the feeder with representative material(s) at a specified feed rate set point. Consecutive timed samples (e.g., 1 min) are obtained and weighed on a precision balance; the weights are converted to feed rates. Accuracy is then defined as percent deviation of the mean feed rate from the feed rate set point. Precision is defined as relative standard deviation (RSD) and can be described at 1 or 2 standard deviations. Acceptance criteria must take into account the nature of the materials used for testing, as the accuracy and precision of feeding equipment are primarily dependent on the flowability of the material (10). If a placebo is used for OQ testing, the placebo should be representative of the materials that will actually be used in the feeding equipment.

C. Performance Qualification

The purpose of PQ is "...establishing documented evidence which provides a high degree of assurance that a specific process will consistently produce a product meeting its predetermined specifications and quality attributes" (11). This is typically accomplished through extended time studies or process runs where the system can be challenged at "worst-case" conditions. Worst-case is defined as "...a set of conditions encompassing upper and lower processing limits and circumstances, including those within standard operating procedures, which pose the greatest chance of process or product failure when compared to ideal conditions (such conditions do not necessarily induce product or process failure)." For a melt extrusion process, those conditions include extremes of feed rate, screw speed, barrel temperatures, as well as possibly including the screw design.

Performance Qualification typically includes a minimum of three valid consecutive runs. The length of time required for each run can vary from hours (e.g., kilograms) to days (e.g., full batch), using either placebo or active compound. The extrusion process is a continuous process and reaches equilibrium within the first few minutes. Samples and data obtained at frequent intervals can be used to verify process stability and consistency.

Acceptance criteria should be specified for both extrusion response parameters (extruder torque, melt pressure, melt temperature) as well as for product performance (e.g., dissolution profile, percent crystallinity, etc.). The outcome of PQ is for the product to meet the acceptance criteria at the specified extrusion operating conditions. Different lots of raw materials should also be included as part of the PQ.

D. Computer Validation

The trend for melt extrusion equipment is to utilize computers to control the process and/or record process data. The relatively small volume of an extruder, coupled with an electronic batch record, provides unique capabilities that are not possible with a traditional batch process. As an example:

1. Assume a melt extrusion process with a 600-kg batch size.
2. Assume an extruder capacity of 12 kg/hr (200 g/min). The batch will take 50 hr of continuous operation for completion.
3. Assume a nominal mean residence time of approximately 1 min within the extruder (typical value). The volume of material contained within the extruder at any given moment is approximately 200 g (200 g/min feed rate and nominal residence time of 1 min).
4. A computer-based data acquisition system could record all process data approximately every 15 sec (four times per minute, or four times per 200 g).
5. The process is documented (feed rate, screw speed, temperatures, pressure, torque, etc.) for every 50 g of material throughout the 600-kg batch.

Electronic control and data acquisition systems are common in nonpharmaceutical extrusion applications where validation is not required. These systems are based on industry-standard hardware and software with application-specific software written as a "front end" for screen graphics. There are no requirements for audits, change control, documentation, etc. in the polymer processing industry.

The recent requirements for electronic documentation systems to comply with 21 CFR Part 11 can involve significant validation effort (12,13). The responsibility can be placed on the supplier to provide a system that is 21 CFR Part 11-compliant, thereby reducing the problems with validating noncompliant systems (14). In the end, it is the responsibility of the user to ensure compliance of electronic systems.

V. CONCLUSION

As more companies explore the benefits of melt extrusion technology, it is becoming more accepted as a pharmaceutical unit operation. This is evidenced by the increase in technical publications and the issuance of patents over the past 5 years. As a result, the extrusion equipment suppliers are responding to these new applications with improved designs and documentation.

From a regulatory perspective, there is no difference between an extruder and any other piece of process equipment. It is the users of this equipment who need to understand what is unique about an extruder as far as validation is concerned. Most of the challenges and issues are cultural rather than technical and require an industry based on batch processing to embrace a continuous technology. As with any other technology, well-written procedures (together with a documented training program) are needed for validating, operating, cleaning, calibrating, and maintaining extrusion equipment and process(es). Well-written procedures require an in-depth knowledge of materials, process, and equipment as well as the interaction between them. It will take time for the pharmaceutical industry to develop the same type of fundamental understanding for melt extrusion as currently exists for other technologies (compressing, granulating, etc.).

REFERENCES

1. Grünhagen HH, Müller O. Melt extrusion technology. Pharm Manuf Int 1995; 167–170.
2. Wood C. Commissioning and qualification: the ISPE Baseline® Guide. Pharm Eng 2001; 2:50–54.
3. Signore AA. Good commissioning practices: strategic opportunities for pharmaceutical manufacturing. Pharm Eng 1999; 3:56–68.
4. Angelucci LA III. Validation and commissioning. Pharm Eng 1998; 1:40–44.

5. Food and Drug Administration. Code of federal regulations. 21 CFR Parts 210 and 211. Fed Regist 1993.

6. Cantwell J. Validation protocols. Pharm Eng 1999; 4:46–49.

7. Saxton B. Reasons, regulations, and rules: a guide to the Validation Master Plan (VMP). Pharm Eng 2001; 3:18–26.

8. Lange BH. GMP manufacturing equipment purchase and qualification: an integrated approach. Pharm Eng 1997; 1:18–24.

9. Roberge MG. Factory Acceptance Testing (FAT) of pharmaceutical equipment. Pharm Eng 2000; 6:8–16.

10. Wilson D. Feeding Technology for Plastics Processing. Munich: Carl Hanser Verlag, 1998.

11. Amer G. An overview of process validation (PV). Pharm Eng 2000; 5:62–76.

12. Anon. Electronic documentation. Pharm Eng 2000; 5:96–99.

13. 21 CFR 11. Electronic records, electronic signatures final rule. Fed Regist 1997; 62:13430–13466.

14. Picot VS, McDowall RD. Containing the 21 CFR 11 problem: purchase of non-compliant systems. Am Pharm Rev 2001; 1:91–96.

18
Controls and Instrumentation

Stuart J. Kapp
American Leistritz Extruder Corporation, Somerville, New Jersey, USA

Pete A. Palmer
Wolock & Lott Transmission Equipment Corporation, North Branch, New Jersey, USA

I. INTRODUCTION

The extrusion control system is referred to as not only the operator/machine interface and monitoring center, but also all of the individual motors, drives, temperature controllers, and measurement devices required to manage the extrusion process. A precise control (and monitoring) of process variables provides the ability to produce a product with repeatable accuracy, high productivity, and low waste. At the same time, the accurate control and monitoring of the extrusion process are extremely important in troubleshooting, and for documentation and quality control purposes.

Typical variables that need to be *controlled* in pharmaceutical extrusions are feed rates of the individual ingredients (feedstreams), barrel/die (zone) temperatures, motor speed, and, occasionally, discharge pressure (Figure 1).

Variables and parameters that need to be *measured and monitored* during the extrusion process include screw speed, set vs. actual temperatures, motor load, "melt" pressure, "melt" temperature, and vacuum level.

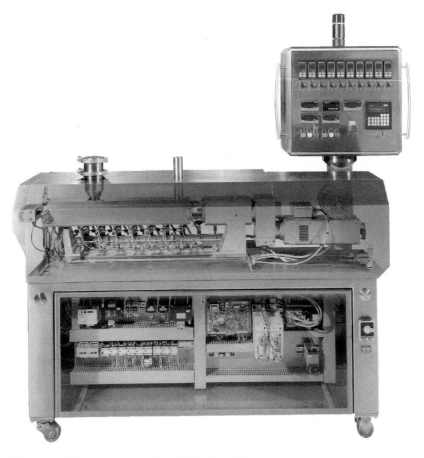

Figure 1   Twin-screw extruder with "onboard" controls.

II.   EXTRUDER TEMPERATURE CONTROL

An accurate temperature control of the extruder barrel, die, and auxiliary equipment is important for any application, but it is even more critical for pharmaceutical applications. External heating is required to bring the extruder and the die up to its operating temperature, and to maintain that temperature for the duration of the process. During continuous operation, most extruders impart large amounts of energy at the screw/material/barrel interface, resulting in considerable heat rise. This mechanical energy input makes sufficient cooling of the extruder just as important as heating.

Extruders utilize one of three methods of heating: electrical, fluid, or steam. While steam was commonly used to heat extruders in the mid 20th century, steam heating is rarely used in extrusion applications today, let alone pharmaceutical applications. Most of today's extruders are heated by means of high-wattage electrical heaters. A limited number of applications, particularly those that run at low temperatures (0–50°C), utilize liquid temperature control.

Cooling, when required, is accomplished through controlled quantities of either air or fluid. Heat transfer fluids common to extrusion are oil (synthetic or hydrocarbon-based), water, glycol, or a mixture of water and glycol.

A. Electrical Heating

Electrical heating is, by far, the most common method of heating modern extrusion machinery, including a vast majority of those machines processing pharmaceuticals. Electrical heating offers numerous advantages over liquid or steam heating, including low cost, a simplistic design, a wide range of temperature control, and decreased maintenance. Electrical heaters, whether they are cartridge-type or "band"-type, are placed along the length of the extruder barrel and the die. These heaters are arranged in groups called zones. Each "zone" of the extruder is independently controlled. Smaller extruders may have as few as two or three zones, while larger, longer extruders may utilize as many as 10–15 (or more) zones (Figure 2).

Electrical resistance heaters are based on the principle of heat generation as a result of current passing through a conductor. The amount of heat generated by any resistance-type heater is determined by the resistance of that heater, and the current flow through it. Accordingly, it is the responsibility of the microprocessor-based temperature controllers to precisely control current flow.

B. Liquid Heating

Heating by means of liquid media offers distinct advantages in its uniformity, and its absence of localized overheating or "hot spots." For these reasons, liquid heating is sometimes preferred for heat-sensitive formulations. Conversely, using heat transfer fluids as a method of heating has considerable drawbacks, including high temperature limitations, possibility of leakage,

Figure 2 Multiple extruder barrel temperature zones (note supply and return piping for water cooling on each zone).

costly periodical disposal of heat transfer oil, and an overall increase in maintenance.

C. Air Cooling

The cooling of the extruder barrel can be accomplished by means of air, water, or oil. Although some dies require extremely tight temperature control and are cooled, this is the exception rather than the norm. While some very small machines may use compressed air through a regulator and solenoid valve arrangement, most air-cooled machines utilize onboard cooling fans, which are activated by their respective temperature controllers (Figure 3). In either case, air is forced through jackets, or across fins that surround the individual barrel zones. More common in today's modern production-scale extruders, and especially in more critical pharmaceutical applications, is the use of water as a cooling media.

D. Liquid Cooling

In addition to water, liquid cooling systems may utilize oil, glycol, or synthetic heat transfer fluids. In most cases, a mixture of water/glycol offers the best solution, as most heat transfer glycols are formulated with special additives to minimize mineral deposits and scaling commonly found in tap water. These additives prevent cooling lines, bores, passages, valves, and associated hardware from prematurely clogging and negatively effecting heat removal.

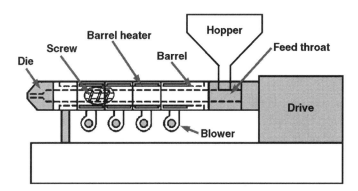

Figure 3 Extruder with barrel heaters and blowers for air cooling.

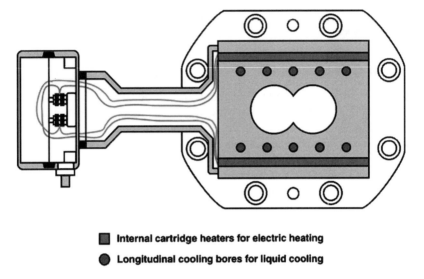

■ Internal cartridge heaters for electric heating

● Longitudinal cooling bores for liquid cooling

Figure 4   Cross section of twin-screw extruder barrel with internal cooling bores.

Liquid cooled extruders circulate the media through passages surrounding the barrel, including the feed throat (which is often cooled only). These passages may be internally cored in the shell of the extruder barrel, or in simple, externally mounted bands or jackets (Figure 4). A closed-loop system, consisting of a pump, a filter, and a heat exchanger, is commonly used to provide a constant supply of tempered water to the extruder barrels (Figure 5). Alternately, and typically at a higher cost, an air-cooled refrigerated chiller may be used for the same purpose, thereby alleviating the need for a city or tower water supply to remove the process heat from the heat exchanger.

E.   Thermocouples

Zone temperature is monitored by means of thermocouples or resistance temperature detectors (RTDs). Thermocouples are widely used as a result of their reliability, small size, low cost, and ease of use. Thermocouples use very low voltage to transmit their signals. Thermocouples are constructed of two dissimilar metal wires with differing thermoelectric behavior. When a variation in temperature is sensed, a small voltage is produced. The level of this millivolt signal is directly related to the temperature of the junction, and thus is a very reliable method of measuring temperature.

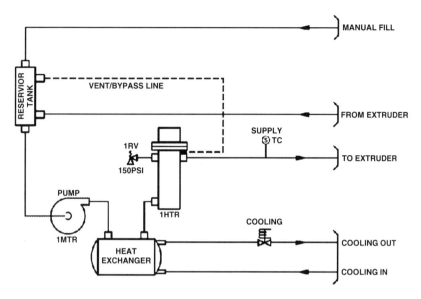

Figure 5   Closed-loop heat exchanger.

Thermocouples may be constructed of numerous different metals, each with its own unique thermoelectric behavior. Accepted types of thermocouples have been standardized by organizations such as the American National Standards Institute (ANSI). Each type is specified by a letter (i.e., J, K, T, E). J-type thermocouples are most commonly used in extrusion applications, while K-type thermocouples and RTDs may be utilized for higher temperature or more demanding applications (Figure 6).

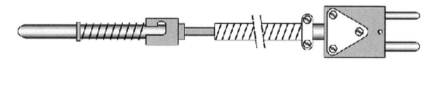

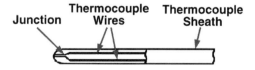

Figure 6   Type "J" control thermocouple.

F. Temperature Controllers

Thermocouple signals are interpreted by temperature controllers, which compare the actual measured temperature against the target set point. These temperature controllers then calculate the appropriate output required to achieve the desired set point. Most of today's temperature controllers utilize microprocessor-based proportional–integral–derivative (P–I–D) algorithms to precisely control extruder temperature (Figure 7).

While these microprocessor-based controllers have numerous operator-selectable parameters, the three main parameters that determine the proper operation of a temperature controller are the proportional band, the integral band, and the derivative band.

The proportional band is a band or range of temperature over which power is proportioned or reduced, as the set point is approached. This allows for a continuous adjustment of the output power (heating or cooling) depending on the actual temperature at any given time. The proportional band is adjustable to yield a stable temperature control over a wide range of process

Figure 7   Extruder control panel with 18 PID temperature controllers.

conditions. The proportional band is typically specified in terms of percentage of instrument span.

The integral band is used to "automatically reset" or continuously adjust the level of the proportional band until the deviation is zero. Once this is achieved, the integrator holds the correct amount of "reset" until the process heat requirements change, or until the temperature is disrupted.

The derivative band helps to prevent overshooting of the temperature set point by compensating for the rate of change. This feature is an anticipatory function that adjusts the output, in an attempt to predict heating/cooling needs. Derivative control is critical in larger machines, where temperature changes occur slowly. Smaller extruders, on the other hand, react much more rapidly to temperature changes and are less dependent on the derivative band.

III. EXTRUDER SPEED CONTROL

Electronic drives control the speed, torque, and direction of either an alternating current (AC) motor or direct current (DC) motor. The basic function of a DC or AC motor is to convert electrical energy and power into mechanical energy and power. Motors operate through the interaction of magnetic flux and electrical current. When electrical current, either AC or DC, flows through coils, it creates electromagnetic energy; the resulting attractive and repelling forces between the rotor and the stator cause the motor to spin.

A. Direct Current Drives and Motors

Direct current drives are used to control the speed and the torque of DC motors by means of a controlled bridge rectifier. Direct current drives convert AC line voltage into variable DC voltage with a silicon-controlled rectifier (SCR) phase-controlled bridge rectifier to power the DC motor armature (Figure 8). A separate field supply from the same drive provides the motor with DC field excitation. Direct current motor speed is proportional to armature voltage and the motor torque is proportional to armature current. An SCR or thyristor is a power semiconductor device with a gate, where the device can be triggered into conduction by means of a firing pulse. This allows the drive to provide an adjustable voltage and current to the motor armature so that the drive can precisely control the speed or torque of the motor under differing load and temperature conditions.

Direct current motors and drives have been the workhorse for the extrusion industry since the 1960s. Early analog drives were set up and tuned

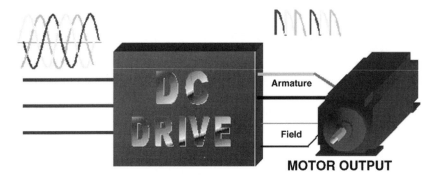

Figure 8   The DC drives convert AC line voltage into variable DC voltage to power the DC motor armature.

with onboard jumpers and potentiometers. Power circuits have changed little since the early introduction of the six pulse thyristor bridge.

In its simplest form, a DC motor consists of a frame and a rotor. The frame or stator is a nonrotating component, and acts as a support as well as provides a flux path for the field. The rotor of a DC motor is called the armature. The armature consists of windings and a commutator fixed to a shaft. The physical size and the weight of a DC motor are directly related to torque and not to horsepower (hp). The bigger and heavier the motor, the more torque it can produce at the shaft. Small DC motors are less efficient than larger DC motors at 85% and 95%, respectively.

There are four basic types of DC motors.

The most popular type of DC motor is the shunt-wound type. A shunt-wound DC motor utilizes an armature connected in parallel across the field windings. A second type, compound-wound DC motors, is similar to the shunt-wound type but includes a series field in series with the armature. Compound-wound motors are typically seen in larger horsepower applications where the motor does not regenerate. These motors offer superior starting torque as compared to shunt-wound motors.

Series-wound DC motors make up yet a third type. These motors utilize an armature and field in series with each other. While these motors offer excellent starting torque, their speed regulation is relatively poor, and therefore would not be a good choice for pharmaceutical applications.

The fourth type of DC motor is the permanent magnet type. Permanent magnet motors are similar to shunt-wound motors but replace the field windings with permanent magnets. Permanent magnet motors are seen mostly in smaller horsepower motors.

Direct current motors are encased in protective enclosures. The basic protective enclosures for DC motors are Open Drip Proof (ODP), Totally Enclosed Non-Ventilated (TENV), and Totally Enclosed Fan Cooled (TEFC). The ODP motors can be force-ventilated with an externally mounted AC blower, or, alternately, may have air ducted in from a separate source. The constant supply of cooling air is independent of motor shaft speed. For this reason, ODP motors can provide full torque to very slow speeds.

The TENV motors have no ventilation. They utilize an internal fan connected directly to the motor shaft to prevent overheating. The TENV motors dissipate heat through the frame of the motor, thereby requiring a larger frame as compared with an ODP motor of equivalent horsepower rating. These frames are usually seen in the lower horsepower range. Totally enclosed motors can be provided in explosion proof construction. Totally enclosed motors are also good for environments that are excessively dusty or dirty, or where it is not desirable to have carbon dust from the brushes in the area.

The TEFC motors are similar to TENV motors but include an external fan connected directly to the main shaft. The fan circulates air over the outside frame of the motor. Because the fan is connected directly to the shaft, it rotates at the same speed as the motor, resulting in poor cooling at low speeds. These motors are typically available in higher horsepower ratings than TENV motors.

Speed regulation is defined as the percent error a motor will have from its base speed if there was a 95% change in load. Speed regulation is dependent upon the accuracy of a speed feedback device. There are three types of speed feedback devices: open-loop or armature voltage feedback (2–3% speed regulation), closed-loop tachometer (0.1–0.2% speed regulation), and closed-loop encoder (0.1–0.2% speed regulation). Note that a closed-loop system refers to a feedback device (Tach or encoder), which monitors the actual speed of the motor and compares it to the reference signal.

B. Alternating Current Drives and Motors

Alternating current drives control speed (and sometimes torque) by converting AC to DC and then back to variable AC. Alternating current drives convert *fixed* voltage and frequency into *variable* voltage and frequency, in order to run three phase induction motors (Figure 9). Inside small AC inverter drives, power is rectified to DC using uncontrolled rectifiers. Larger AC drives rely on SCRs to perform the same task. The DC power is then filtered, and converted back to AC using solid state transistor switches.

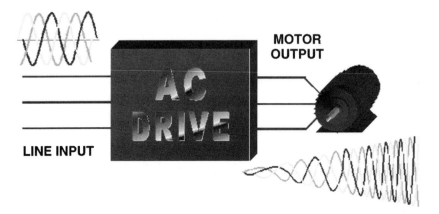

Figure 9   The AC drives convert the "fixed" voltage and frequency into "variable" voltage and frequency to run three-phase induction motors.

Alternating current drives operate by varying the frequency and voltage to an AC motor. The frequency of the applied power to an AC motor determines the motor speed. To maintain constant torque, the drive must maintain a constant voltage/hertz ratio. Pulse width-modulated (PWM) drive techniques (to control the inverter bridge) are, by far, the most popular technology for providing a simulated variable sine wave to the motor. The PWM-type drives have historically required a more complex regulator than previous designs; however, the use of today's high-power microprocessors has eliminated all but this problem. The microprocessor and other recent technological advances in digital AC variable frequency drives now allow them to support applications where DC drive technology was traditionally the choice.

There are three basic types of AC drives: open-loop voltage/ hertz drives, open-loop vector or sensorless vector drives, and flux vector or closed-loop vector drives

1.   Open-loop voltage/hertz drives: This is the most basic type of AC drive. Open-loop voltage/hertz drives are open loop, which means there are no speed feedback devices. Motor voltage is varied linearly with frequency. Most voltage/hertz drives cannot separate torque-producing current from flux-producing current, so the voltage/hertz drive cannot regulate torque. The main drawback with voltage/hertz drives is that they cannot compensate for motor and load dynamics, resulting in poor speed regulation. Voltage/hertz drives also lack low speed/starting torque. The advantages of voltage/hertz drives include ease of set up, low cost, and lack of required feedback devices.

2. Open-loop vector or sensorless vector: This type of drive is also open loop and does not require an encoder or tachometer to feed motor data back to the drive. Instead, open-loop vector drives utilize an advanced algorithm that creates a mathematical model of the motor's electrical characteristics to provide feedback within the drive. This configuration allows the drive to respond to sudden load changes by calling for more or less torque-producing current. Although sensorless vector drives yield higher low-speed torque, they lack the dynamic response or high-performance speed regulation of the flux vector drive.

3. Flux vector or closed-loop vector: This drive does require a motor-mounted feedback device to supply information to the drive on the rotor's position relative to the stator. The flux vector drive can control both the flux-producing current and the torque-producing current of the motor. This allows the drive to regulate both speed and torque. This drive can provide full torque at zero speed and most can provide 150% starting torque. The flux vector drive has excellent shock load response characteristics and high-performance speed regulation.

Many of today's drives can provide all three modes of operation in one drive. Changing from one mode to another mode would be a single software parameter.

The AC "squirrel cage" induction motor is the fundamental workhorse of industry. The basic parts of a three-phase AC induction motor consist of a rotor, a stator, and two end shields housing the bearings that support the rotor shaft. The stator is the nonrotating component that contains the three phase windings. The revolving section of the motor is the rotor containing steel laminations around the motor shaft. Alternating current power is supplied directly to the stator and produces a rotating magnetic field. The motion of the magnetic field induces a magnetic field in the rotor. The induced field in the rotor rotates with the magnetic field in the stator and causes the motor to spin.

Early AC inverter drives were connected to regular AC induction motors. Alternating current PWM inverters can yield very high switching frequencies and very rapid changes in voltage that commonly result in pinholes in the insulation, causing short circuits and premature motor failure. Many motor manufacturers introduced lines of motors they call inverter duty or vector duty motors. High dielectric strength wire insulation is typically found in these motors to resist such pinholes. Cooling performance can be increased by simply adding constant speed blowers, or by using oversized frames in nonventilated motors. Alternating current motors allow for feedback sensing and wider speed ranges.

Much of the heat generated by an AC motor is in stator. The outside frame of the motor is designed with fins to help dissipate the heat to the immediate atmosphere. To minimize frame size, constant speed blowers are attached to the motor. Larger horsepower AC motors can be manufactured in drip-proof force-ventilated enclosures to provide forced cooling air through the motor.

C. Digital Drives

Since the 1980s, microprocessors have enabled the rapid development of cost-effective digital drives. The evolution of digital drives has brought the following advantages: improved reliability through a reduction in the number of components, elimination of drift problems associated with analog values, extreme levels of accuracy, excellent repeatability, digital communications, increased monitoring and diagnostics, complex algorithms, and ability to autotune a drive (match motor to control). Microprocessor-based controls now allow drive parameters to be downloaded to and from a computer for backup purposes using personal computer (PC) drive configuration software.

Most digital drives have the ability to serially communicate, typically through an add-on board. The microprocessor has made possible the signal handling functions required. Drives can now be controlled and adjusted remotely with a complete set of information being returned for analysis. One major advantage of this serial communications is that a programmable logic controller (PLC) or computer can communicate to many different drives and devices with just one daisy chain or multidrop connection. Each device on the communications network is given a unique serial address.

There are many different types of industrial network technologies used with digital drives. Some are proprietary and some are open networks. Open networks are accessible to anyone or any manufacturer that wants to make a product "talk" on that network. The different networks differentiate themselves from one another by the number of nodes allowed, maximum cable length, data rate, network model, and performance requirements.

Serially communicating to drives and devices brings numerous advantages. These include a reduction in interconnection wiring [one communications cable would replace most of the hardwired input/output (I/O) to a drive or drives], the ability to display faults, settings and feedback parameters on a central monitor for diagnostical purposes, and the increased accuracy of digital settings. Serial communications also eliminate drift associated with analog devices and inherent inaccuracies related with analog-to-digital converters. Fiber optic communication technologies are immune to noise.

D. Comparison of Motors and Drives

Direct current drives have several advantages of over AC drives. To begin with, DC drives utilize a much simpler controller design, where there is only one power conversion stage and no power storage elements (capacitors). Direct current drives also offer higher controller efficiency. [The efficiency of a system is defined as the ratio of power output (mechanical) to power input (electrical), with the rest being lost as heat.] Losses within the controller are mostly due to the conversion of power or thyristor bridge. Direct current drives are typically 98% efficient.

Direct current drives also offer cost advantages, especially in those sized above 100 hp. They benefit from high controller reliability and low maintenance due to a lone power conversion module.

Other advantages of DC motors include lower power line harmonic contribution, the need for smaller line reactors, compact controller design, better overload and peak voltage characteristics, low motor acoustical noise (AC drives have carrier noise), fewer motor lead–length issues (the distance between AC motor and AC control needs to be minimized to prevent reflected voltage wave), easier troubleshooting, and serviceability.

Alternating current motors have several advantages over DC motors. These include decreased and simplified maintenance (no brushes to replace and no commutator to clean); smaller motor frame sizes than equivalently sized DC motors; high dynamic performance (low rotor inertia compared to the DC armature); small to midsized (1–50 hp) AC motors that are inexpensive and readily available; suitability to harsh, rugged environments (totally enclosed motors go up to high horsepower ranges; can be made explosion-proof or washdown-rated); more precise open-loop speed regulation; wider speed ranges (many smaller standard inverter duty motors can go twice their base speed); longer power-dip, ride-through capabilities due to power storage elements on DC buss; and near-unity power factor regardless of speed or load.

As electronic power components become more compact and reliable, the cost and the size of AC and DC drives will continue to decrease in the future.

IV. PROGRAMMABLE LOGIC CONTROLLER AND PERSONAL COMPUTER CONTROL

Extruder control systems must perform three basic functions: machine control, operator interface, and communications. When manual/discrete controls are

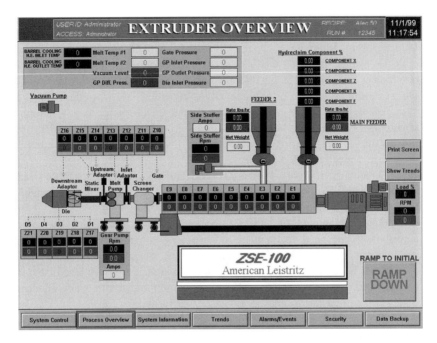

Figure 10   A PC-based operator interface to a PLC control system.

deemed unacceptable, machine builders can either use a PLC, a PC, or a combination of both to perform these control functions (Figure 10).

A.   Programmable Logic Controller Control

The PLCs have been around for the last 30 years. They are still the controller of choice for most automated machineries. They have the ability to process machine logic without experiencing faults in an operating system. They are ideal for critical applications (such as the manufacture of pharmaceuticals) due to their reliability.

Shortly after their introduction, PLCs began replacing relay logic, which was previously used to control machinery. Their advantage over relays was that they were programmable, whereas relays were hardwired. Relay logic required excessive space and the logic could not be easily changed. While early PLCs were also very large, expensive, and only capable of binary logic, their capabilities soon expanded to include analog I/O, thermocouple inputs, and numerous communications options. The ladder logic language used to program PLCs was the same as the ladder logic drawings used for relay wiring and was therefore well understood by electricians, further enhancing their industrial acceptance.

The advantages of PLC-based controls include reliability, low maintenance costs, ruggedness/durability, integral I/O bus design, fast boot times, and large installed base. Additionally, one specific type of visual programming language, Relay Ladder Logic, is easy to learn and understand. The PLC-based controls offer a lower purchase cost for small applications (Micro-PLC), while single sourcing of hardware and software increases reliability and creates a single point of accountability.

B. Personal Computer Control

The PC-based controls utilize a PC acting as the brains of the system. In this scenario, the PC handles not only the human/machine interface (HMI) tasks, data handling, and communications, but it also controls the entire process. The PC-based control option gives us the hardware and programming device in one unit.

The PC-based controls offer the following advantages: large memory and storage capacity, vast availability of commercial software, integral audio/video, complete HMI functionality in one platform, and relative ease of interfacing to other systems for factory data collection. The PC-based controls allow the use of off-the-shelf products from different suppliers without an excessive retraining of personnel. Extensive data handling, as is typically required in pharmaceutical settings, is well suited to a PC-based control system.

The approach of combining the PLC and the PC clearly provides the benefits of both systems. In this scenario, the PLC directly controls the process while the PC handles the HMI and data logging functions. This solution is ideal for applications requiring a high level of machine control interaction by an operator, and for critical process control. This is often the preferred method utilized to comply with stringent regulatory requirements, including those of the Food and Drug Administration (FDA).

V. PROCESS MONITORING

While it is the function of the individual components (PLC's temperature controllers, drives, motors, thermocouples) to precisely control their respective extruder functions, it is imperative that data are fed back to the operator to verify that the overall process is running as expected, and within acceptable limits.

Basic, discrete (digital) indication of speed, temperature, pressure, and motor load is standard on almost all extruders, and may be sufficient for noncritical applications and processes. Pharmaceutical processes, however, typically require constant monitoring and recording of these variables in order to satisfy stringent documentation and accountability requirements. In this case, a PC is often employed to handle the tedious task of extensive data collection, organization, and manipulation. The four critical parameters that need to be monitored in any extrusion process are screw speed, motor load, temperature(s), and pressure(s).

A. Screw Speed

Screw speed is a critical process variable, in that it often dictates the amount of energy, the degree of mixing, the melt temperature, and sometimes the melt pressure of the extrusion process. In single-screw extrusion, screw speed also dictates the overall throughput, thereby affecting the dimensions of the final product. (This is not true for starve-fed twin-screw extruders, where throughput is determined by feed rate.) While DC motor/drive packages can attain control of accuracy of anywhere from 0.1% to 3%, modern AC motor/drive packages can attain accuracies as good as 0.01% and thus are the preferred choice in pharmaceutical applications. Due to the drastic implications that screw speed has on the overall process, it is imperative that speed be monitored and controlled accurately.

B. Motor Load

Motor load (amperage) is a key indication of how much energy is being transmitted from the motor, through the screw(s), and into the material/ process. A fluctuating, unstable motor load often reveals problems that may otherwise be invisible. An unsteady motor load may be an indication of an inconsistent feed or bridging at the feed throat, or an inadequate speed control due to problems with either the motor or the drive. Conversely, a steady, stable motor load typically indicates a steady, controlled process. As with temperature and pressure, it is best to graphically plot motor load over time, as simple, instantaneous "snap shots" of the motor load will not reveal significant trends or shifts in amperage.

C. Melt Pressure

The pressure generation at the tips of the screws (also referred to as "head pressure") is a critical process parameter that should be monitored carefully and accurately. Some extruders will also position pressure transducers directly in the die. Variations of melt pressure can indicate process interrupts or instability, such as inconsistent feeding, melting, or throughput. Conversely, a steady, uniform pressure reading typically indicates a stable extrusion process. The accurate monitoring of an extruder's head pressure is also an important safety measure, as overpressurizing an extruder barrel can lead to serious operator injury. For this reason, most extruders are "interlocked" to the pressure indicator, so as to automatically shut down the extruder in the event a given pressure (operator selectable) is exceeded for a given period of time.

Modern pressure transducers utilize strain gauges to detect variations in pressure on their diaphragms (Figure 11). Strain gauge-type transducers offer good resolution and quick response. One limiting factor of strain gauges is their susceptibility to high temperatures. For this reason, a hydraulic membrane separates the diaphragm and the strain gauge. Mercury is the most common type of fluid used in these membranes due to their low thermal expansion and high boiling point. However, mercury-filled transducers are strictly prohibited in pharmaceutical applications, as a puncture of the

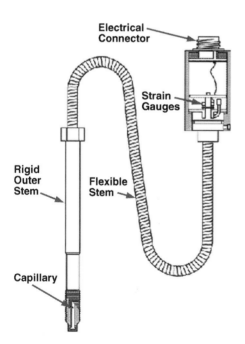

Figure 11   Strain gauge-type pressure transducer.

diaphragm (although extremely unlikely) would be catastrophic. For this reason, sodium-filled transducers are typically utilized for food and pharmaceutical-grade applications.

D. Melt Temperature

Commonly referred to as "stock" temperature, the temperature of the product at any given point in the extruder is of vital importance, particularly as it emerges from the die. As with melt pressure, knowledge of the melt temperature will give an indication of just how stable the process is. In order to precisely measure stock temperature, a thermocouple should protrude directly into the polymer flow. Unfortunately, placing a thermocouple probe directly in the melt stream is usually impractical, as the extremely tight clearances between the screw and the barrel wall prohibit this. Even in areas of the extruder where the use of a protruding thermocouple is possible (head, adapters, die), other negative implications exist, including disruption of the flow orientation, and stagnation or hang-up. The common solution to this problem is the use of a "flush mount" thermocouple probe, where the measuring surface is flushed with the barrel wall or internal die surface. It must be understood that while flush mount thermocouples are only able to detect the melt temperature of the material at the wall, they give a good indication of "apparent" melt temperature—a temperature that can be used to compare similar or identical formulations from run to run, or from lot to lot.

VI. DATA ACQUISITION AND MONITORING SYSTEMS

An accurate recording of machine settings and process data is essential in pharmaceutical settings. Not only do they serve as a basis for comparison for future processes, but they also provide a means for accountability as required by the FDA, the International Organization for Standard (ISO), and other recognized industry standards.

The ability to record key process variables over time reveals important trends and shifts in process conditions that may otherwise go undetected. Prior to the evolution of the PC, strip charts were often used to automatically record machine and process data vs. time. These strip charts were cumbersome, prone to misinterpretation, and required manual tabulation back into numerical data for statistical process control analysis.

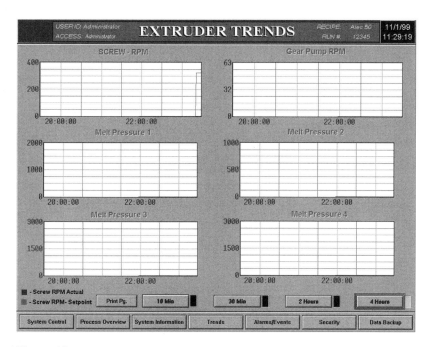

Figure 12 The PC-based data acquisition, monitoring, and trending.

The PC revolutionized the way in which process data are recorded and manipulated. Computer-based data acquisition and monitoring systems allow the operator to instantaneously view the entire extrusion process. All measured parameters are clearly presented to the operator, either numerically or graphically. In addition, any alarms that were triggered during the run are also saved with the file. The speed and the storage capabilities of today's computers allow data points to be collected several times per second, over long periods of time. This information can be stored infinitely for future recall and for comparison against similar extrusion runs (Figure 12).

VII. CONCLUSION

The extrusion control system is the primary link between the operator and the overall machine system. The main purpose of the control system is to provide an efficient means to control and to monitor key process parameters including (but not limited to) speed, temperature, and pressure. While more simplistic extrusion systems utilize DC motors and manual/discrete controls, today's pharmaceutical control systems routinely utilize cutting edge technology including PLCs, AC motors, and drives, and sophisticated PC-based data monitoring and storage systems.

REFERENCES

1. Newtown P. Drives and Servos Yearbook 1990. Control Techniques, 1989.
2. Application Guide Adjustable Frequency Drives and AC Motors. Emerson Industrial Controls, ADG-080B, March 1993.
3. Warner SECO AC/DC Drives Application Notes. Warner Electric, October 1993:1101.
4. Linder N. Choosing the Right AC Drive Type. Eurotherm Drives Regional Seminar.
5. Cooksey L. Driving process control with vector technology. Motion Control Mag 1999; July/August, pp. 24–27.
6. Baldor Motors and Drives Technical Handbook. Fundamentals of Inverter-Fed Motors. Baldor Electric Company, NM780, February 1996.
7. Morley D. The History of the PLC as Told to Howard Hendricks. R. Morley Incorporated, 1996–2001.
8. Gould LS. When Controls Converge: CNC, PLC & PC, Automotive Design & Production, Gardner Publications, Inc., Cincinnati: January 1999.
9. Fisher T. Extrusion Control, The SPE Guide on Extrusion Technology and Troubleshooting. Chapter 4. Society of Plastics Engineers, 2001:1–14.
10. Whelan T, Dunning D. The Dynisco Extrusion Processors. 1st ed. London, England: Dynisco Inc., 1998.
11. Todd D. Plastics Compounding Equipment and Processing. Ohio: Hanser/Gardner Publications, 1998.

19
Future Trends

Isaac Ghebre-Sellassie
MEGA Pharmaceuticals, Asmara, Eritrea, and Pharmaceutical Technology Solutions, Morris Plains, New Jersey, USA

Charles Martin
American Leistritz Extruder Corporation, Somerville, New Jersey, USA

I. INTRODUCTION

It is apparent from the preceding chapters that extruders and high-intensity twin-screw mixers are highly flexible and efficient mixing devices that are suitable for the manufacture of products that demand consistency and superior quality. This requirement is particularly true with pharmaceuticals where the precise delivery of active entities to a specific site within the human body is critical, and variations in delivery could either reduce the effectiveness of the medication, or lead to toxic levels that could put the patient at risk. As a result, it has been the primary objective of the pharmaceutical industry and the regulatory bodies to ensure that drug products distributed to the public are manufactured under Good Manufacturing Practices (GMPs) and consistently meet stringent quality requirements. Because single-screw and twin-screw mixers are equipped with precisely designed and machined hardware and controls, these devices provide the user with manufacturing capabilities that easily meet process validation requirements as mandated by regulatory agencies. Given these desirable attributes, it is expected that single-screw and twin-screw extruders, or variations thereof, will soon become standard pharmaceutical manufacturing equipment, and the application of extrusion processes in pharmaceutical development and production will be firmly established. Extruders do not only provide products that are superior in quality to those manufactured by any other comparable conventional manufacturing equipment, but may also be cost-effective because they allow continuous processing, provided that optimal conditions that exploit the full potential of the devices are met. It is anticipated that significant effort will be devoted to extensive characterization and optimization of formulation variables and processing parameters to accommodate the ever-increasing quality requirements that pharmaceutical drug products must satisfy to gain regulatory approval and marketing authorization. Concurrently, it is expected that equipment manufacturers will continue, with inputs from pharmaceutical scientists and engineers, to further refine machine engineering designs and to build advanced pieces of equipment that keep up with the quality requirements and performance needs of drug products that support a seamless transition from a development environment to a production setting. Some equipment-related design considerations that are likely to be investigated and the impact they will have on dosage form manufacturing are discussed below.

II. EQUIPMENT CONSIDERATIONS

A. Feeders

Solids and liquid feed systems will continue to be modified to accommodate the specific needs of pharmaceutical manufacturing. In particular, feeding devices will be downsized so that very small quantities can be precisely delivered to the extruder system to maintain formulation accuracy. The first development phase will definitely focus on the mechanical designs of these "microfeeders," and will subsequently be applied to the controls technology as warranted. There will also be a trend toward standardization of the controls logic using "off-the-shelf" hardware and software, as compared with customized software developed exclusively by the feeder manufacturers.

B. Single-Screw Extruders

Single-screw extruders are typically utilized as pressure generation devices, and have been used for many years to extrude precision products, such as medical tubing and films for packaging. Drawing upon technologies that have traditionally been used with twin-screw mixing extruders, such as starve feeding and the implementation of extensional shear mixing elements, will allow the single-screw extruder to function as a mixer for some pharmaceutical applications. These devices have not as yet been adequately configured to meet pharmaceutical processing needs, particularly with regard to clean-ability. Specialized manufacturers have recognized this drawback and will continue to modify these highly standardized machines for usage in a GMP environment, which will also result in increased costs for this relatively simple machine.

C. Twin-Screw Extruders

Twin-screw extruders are continuous mass transfer devices that are primarily used for mixing and devolatilization. As compared with the single-screw extruder, this device has been utilized in the pharmaceutical arena for a number of years. It is, however, still in its infancy from a GMP design standpoint. Specifically, segmented designs for screws and barrels that are ideal in a laboratory environment will be replaced in production settings with one-piece barrels and screws that will maintain the same heat transfer characteristics as modular designs. Just as the single-screw extruder will draw upon twin-screw technologies to make it a better mixer, twin-screw machines will be called upon to make precision products that formerly were made on single-screw machines. This will involve the use of longer process sections with screw modifications at the discharge end to build and to stabilize pressure, as well as the integration of front–end pressure-generating devices (i.e., a gear pump) and the use of sophisticated system control algorithms to maintain pressure in a consistent and repeatable way.

D. Metallurgy

For both single-screw or twin-screw extruders, additional efforts will be made to manufacture barrels and screws of metallurgies that are neither reactive, additive, nor absorptive with the materials being processed. The use of hardenable stainless steels is now common, but requires special handling while cleaning. Nickel-based alloys are potentially alternative materials that need to be explored, particularly for corrosive environments. Bimetallic options, especially for barrels, have been highly developed for the plastics industry and will be applied to pharmaceutical processing.

E. Downstream Operations

Just like extruders, die and downstream technologies for pelletization and shape extrusion are highly advanced in other industries, including, but not limited to, the automotive, packaging, electronics, and medical device markets. Existing sophisticated industrial manufacturing methods will be applied on a miniature scale in a GMP format to pharmaceutical processes by specialized machine manufactures that can see the potential of this growing market.

The use of coextrusion for both profiles and films will allow unique drug delivery systems to be manufactured (Figure 1). Another possibility is the injection of a carbon dioxide gas into the process to produce a "foamed" product. It is only necessary to be aware of the incredibly wide array of products that are extruded to envision the possibilities that are available.

Another possibility is the combination of the continuous twin-screw extrusion system with an in-line cyclical molding operation. In this instance, a twin-screw extruder will continuously mix and/or devolatilize a formulation, while filling an accumulator. Downstream from the accumulator, either a screw or a gear pump is intermittently actuated to fill a mold, which will cool/

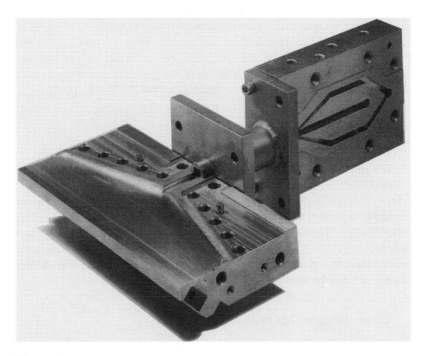

Figure 1   Example of an internal geometry of coextrusion tooling for a transdermal product.

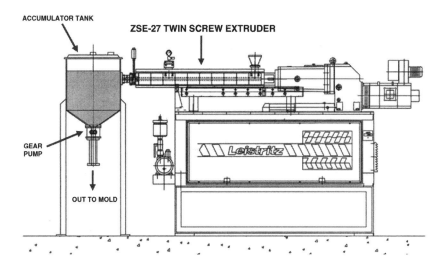

Figure 2 Continuous twin-screw extrusion compounding/devolatilizing cyclic molding system.

size the molten materials into a customized shape, thereby combining the attributes of both a continuous and a discontinuous process in a single manufacturing step (Figure 2).

F. Controls

The same scenario can also be applied to controls. Precision drives and temperature control systems are already prevalent in the industry and merely need to be applied to the pharmaceutical extrusion process. For system and plant integration, the human machine interface (HMI) industrial software is available to adequately control and manage the extrusion system, to acquire data, and to be integrated with the overall manufacturing/plant information system. Control architectures will become, to a significant degree, standardized in the pharmaceutical market. Companies that have already specialized in programming and documentation for other manufacturing operations governed by Food and Drug Administration (FDA) regulations will probably become lead suppliers, as opposed to extruder manufacturers developing this capability in-house. Extruder suppliers will, however, be required to develop partnerships with the controls specialists to facilitate turnkey installations.

III. DOSAGE FORM CONSIDERATIONS

A. Content Uniformity

One of the most challenging quality requirements in pharmaceutical manufacturing is content uniformity of dosage forms. Batch blenders such as V-
blenders, high-shear mixers, bin blenders, etc., which have been employed
traditionally to blend powders that differ in density, particle shape, particle size
distribution, and surface properties, do not consistently produce a uniform
distribution of ingredients. Because the blending volume of these devices is
large, the mass transfer distances are large and the components of the
formulation take a very long time, if at all, to mix uniformly. In contrast to
batch mixers, extruders, particularly intermeshing twin-screw extruders, are
highly efficient mixers that invariably ensure content uniformity of products.
The short mass transfer distances in the small area bounded by the shapes of
the screws and barrel surfaces, coupled with the self-wiping properties of the
system, lead to a highly stable and robust process that provides an extremely
uniform and reproducible product. This is especially true where accurate
distribution of a small constituent of a formulation is desired. The degree of
contribution of these components is directly related to the screw arrangement
on the shafts and, to a large extent, to screw design. Both distributive mixing,
where particles are spaced within the system, and dispersive mixing, where
particles are broken down into smaller units up to and including the molecular
level, are key elements of the totality of the mixing process. Most of the screw
designs available today were designed to address the needs of the plastics and
rubber industry. While these elements have significantly improved the content
uniformity of pharmaceutical dosage forms compared to traditional pharmaceutical manufacturing processes, these are by no means ideal. As the
utilization of extruders gains acceptance in the pharmaceutical industry, it is
likely that new screw elements that address the low viscosity of most
pharmaceutical formulations and further improve the quality requirements of
drug products will be designed in the future.

B. Dosage Form Flexibility

Extrusion, along with its downstream operations, provides the development
scientist with opportunities to tailor-make various shapes of oral dosage forms.
Some of these have already been demonstrated to be commercially feasible
and include injection molding (Zetachron) and calendering (Knoll Soliqs)
processes. It is likely that variations of these technologies will appear in the

future. Coextrusion is another extrusion process that has a great potential in the development of a variety of controlled-release dosage forms practically impossible to manufacture by conventional processes. The technology is applicable to the production of single units, multiparticulates, and multi-laminate films. Another emerging technology that has had some applications in the food industry and could have extensive application in the pharmaceutical industry is supercritical fluid/extrusion. Supercritical fluid/extrusion processes, where a supercritical fluid (carbon dioxide gas) is injected into a formulation processed in an extruder, have the potential to produce highly porous matrices that are suitable for the development of fast-dissolving solid dosage forms (1–3). Fast-dissolving oral dosage forms are becoming increasingly popular not only for the development of fast-dissolving/fast-melting products suitable for pediatric and geriatric applications but also for those drug products that require a fast onset of action. It is likely therefore that supercritical fluid/extrusion will become a widely used manufacturing process for these specialized dosage forms.

C. Excipients

It would have been desirable to manufacture pharmaceutical solid oral dosage forms that contains only the active ingredient. Instead, active ingredients need to be intentionally adulterated with various excipients that have different functions before drug products are manufactured. Some excipients are incorporated to facilitate manufacturing, while others are added to enhance the dissolutions and hence the bioavailabilities of drug products. As a result, pharmaceutical dosage forms contain active ingredients and a number of other additives, the qualitative and quantitative compositions of which depend on the type of dosage form and/or the manufacturing processes. In general, traditional blending processes are highly inefficient and require formulations that contain excessive amounts of excipients for a particular desired function. These inefficiencies lead to relatively higher cost of goods and, more importantly, to dosage forms that are larger than needed for oral administration. Extrusion, by virtue of its inherent intense mixing capabilities, allows an efficient utilization of these excipients, thereby significantly reducing the cost of goods and providing smaller dosage forms suitable for swallowing. As screw elements designed for pharmaceutical manufacturing are refined, the mixing efficiency of extruders will be enhanced, further improving excipient utilization. Given these benefits, it is expected that extrusion will be preferred techniques for the mixing and melting of pharmaceuticals.

D. Manufacturing Steps

It has been stated in previous chapters that extrusion processes can handle multiple operations through screw design configurations and upstream and downstream operations. These operations may include conveying, melting, mixing, shaping, and cooling during melt extrusion; and mixing, granulation, wet milling, and drying during wet granulation. The blending of the formulation components is a separate step carried out before the extrusion process to ensure that the material fed into the extruder is roughly uniform before it is fed into the extruder. Otherwise, in spite of its mixing capabilities, the extruder cannot bring about uniformity in a product when a formulation that enters the small mixing volume between the screw flights and the barrels is not a representative sample of the total mass. The short mass transfer distances within the extruder work better when the individual ingredients are fed into the extruder accurately. Currently, feeder technology does not allow the feeding of separate pharmaceutical powders in a formulation to be fed at precisely the desired rates. However, as screw designs are refined further and the feeders technology is advanced, individual excipients, including minor components, will be added directly into an extruder, eliminating the blending step from the manufacturing line. Coupling the extruder with a downstream operation like injection molding, calendering, etc. can also reduce the manufacturing steps further. All these technologies are being explored and will be used more as the industry becomes comfortable with extrusion technology.

E. Scalability

One of the most serious challenges to pharmaceutical manufacturing has been the scale-up of batch processes from a development environment to a production setting. Scale-up issues have been responsible for numerous delays in the introduction of drug products into the market, thereby costing companies millions of dollars in revenues. While significant progress has been made in understanding the critical parameters of batch processes, scale-up will always be an issue for these processes. Mass transfer distances within these processes, particularly the mixing and blending operations, differ significantly as the processes are scaled up from small vessels to large blenders and granulators. These differences, coupled with differences in energy input, make the scale-up process unpredictable, often times requiring multiple bioequivalency studies to assure the sameness of products. Extrusion processes, on the other hand, are better suited. In most cases, there may not be a need for scaling up from small-size equipment to production-size equipment. The processes, being continuous, only require longer run times using the same equipment and process parameters to increase the batch sizes. This permits drug products manufactured under the same conditions to be used during clinical efficacy studies and production batches without the need for multiple bioequivalency studies. In situations where scale-up is required, the parameters that govern the scale-up of extrusion processes are better understood, and the probability of achieving bioequivalency between clinical and commercial batches is high. This attractive feature is expected to increase the demand for extrusion processes in pharmaceutical manufacturing in the future.

F. Automation

Another important attribute of extrusion is the suitability of the process for automation. Batch processes, the most prevalent processes in the pharmaceutical industry nowadays, are not only labor-intensive but are also prone to human error, thereby affecting the quality aspects of drug products. It is widely believed that automation improves quality and brings about manufacturing efficiency. As such, the pharmaceutical industry is constantly trying to automate manufacturing processes whenever possible. Because extrusion has the potential to link all the pharmaceutical unit operations into a single manufacturing line, up to and including the packaging step, it is expected that the industry will resort to extrusion and its downstream operations to improve quality and to cut cost through automation. This is especially critical when one considers that manufacturing costs now generally make up 18–20% of a company's operating costs, which is almost twice what it was a decade or so ago (4).

IV. CONCLUSION

It is apparent that extrusion provides the pharmaceutical industry with significant opportunities to improve upon the manufacturing technologies in current use. As companies feel comfortable with extrusion processes and the relevant regulatory requirements, more of them will embrace the technology. Some have already done so and others will follow. Engineering firms will also target more resources to build market specialized extrusion machines for use in a pharmaceutical environment. In addition, it is expected that for those companies that do not want to invest in capital equipment early on, contract manufacturing laboratories will be established and become more readily available for feasibility studies when a product is in a preliminary stage.

REFERENCES

1. Alavi SH, et al. Structural properties of protein stabilized starch-based SCF extrudates. Food Res Int 1999; 32:107–118.
2. Breitenbach J and Baumgartl H. Solid foamed active substance preparations. U.S. Patent 6,150,424.
3. Elkovich MD, et al. Effect of SCF CO_2 on morphology development during polymer blending. Polym Eng Sci 2000; 40(8):1850–1861.
4. Bergsten C. Collaborative approaches streamline capital project delivery. Pharm Manuf Fall 2002; 11–12.

Index

Abrasive wear, 22
Acceptance criteria, 356–357
Accuracy, 156, 356
AcDiSol, 189
Adhesive wear, 22
Aerated solid, 120
Agglomeration, 88, 324, 327
Agitator, 123
Air chuck, 234
Air jet mill, 262
Air knife, 230, 234
Air quench pelletizing, 174
Air rack, 212
Air ring, 228
Air stripper, 172
Alarms/Interlock, 355
Alternating current motor, 369, 371, 375
Amorphous precipitation, 246
Amorphous, 265
Annular die, 103
Antioxidant, 194
Apex, 71, 86
Archimedean transport, 49
Arching, 41
Ascorbic acid, 194

Aspirin, 262
Atomizer, 174
Autohesion, 307
Automation, 391
Axial extruder, 289, 291
Axial single-screw extruder, 297
Axial, twin-screw extruder, 298

Ball mill, 262
Barrel cooling, 22, 347
Barrel friction, 46
Barrel heating, 22
Barrel temperature, 160
Barrel, 20, 158, 163
Barrier mixer, 65, 145
Batch, 324
Batch compounder, 137
Batch mixer (*see* Batch mixing)
Batch-mixing, 140–141, 286
Batch process, 316, 324, 390–391
Batch record, 351
Belt conveyor, 212
Belt feeder, 155
Belt puller, 216
Biaxially orient, 230
Bilobal, 86

Bioavailability, 245, 249, 261–262, 264, 267, 274
Bioequivalence, 255
Blending volume, 388
Blown film, 228
Boundary condition, 70
Breaker plate, 40
Buccal drug delivery, 200
Bulk deformation, 99
Butylated hydroxytoluene, 194

Calendering, 253, 388, 390
Calibration tooling, 211
Capillary die, 100, 150
Capillary rheometer, 150
Capillary, 150, 284
Carnauba wax, 189
Cascade system, 316
Casein, 264
Cast film, 229
Cellulose acetate butyrate, 188
Cellulose acetate phthalate, 189
Cesamet, 246
CFR 21, Part 211, 348
CFR Part 11, 358
cGMP, 345
Channel volume, 55
Channeling, 106
Characterization, 135, 137
Chill roll, 229
Chlorpheniramine maleate, 195, 304
Choker bar, 238
Chrome plating, 231
Circulating coolant, 351
Citrate ester, 188
Clamshell-design barrel, 148, 353
Cleaning, 29, 30
CMP consideration, 243
Coextrusion, 99, 386, 389
Coil winder, 221
Co-kneader, 25
Commissioning, 345, 347–348
Compaction, 279, 312

Compounding, 141, 145
Compression, 186, 193
Computer validation, 357
Computer-aided engineering, 99
Conical twin-screw system, 147
Constrained flow, 99
Content uniformity, 167, 193, 195, 388
Continuous cutting, 219
Continuous granulation, 336
Continuous mixer, 287
Control algorithm, 228
Control parameter, 20, 352
Controlled release, 183, 187, 278–279, 312
Convection, 231
Conveying zone, 186
Cooling belt, 173
Cooling bore, 23
Cooling medium, 210
Cooling table, 212
Cooling tank, 213
Coriolis mass flow meter, 118
Corotating twin-screw extruder, 8–9, 69, 71, 96, 154, 186
Corrosion, 22
Counterrotating twin-screw extruder, 7–9, 28, 69, 71, 162, 186,
Crammer feeder 50
Cross-hatched pattern, 298
Cutter bushing, 218
Cyclical molding operation, 386
Cyclodextrin, 250
Cylinder extruder, 291, 296

D-α-tocopheryl polyethylene glycol 1000 succinate, 194
Danazol, 264
Data acquisition, 380
Decompression zone, 58
Deformation, 279
Densification, 326–327
Derivative band, 369
Design qualification, 349

Devolatilization, 33, 35, 92–93
Die face pelletizer, 174
Die hole freezeoff, 176
Die length, 295
Die lip, 102
Die pressurization, 162
Die swell, 179
Die, 99, 103, 161–162, 179, 184, 186, 209, 228–229, 284, 289, 291–292
Diethyl phthalate, 188
Digital drive, 374
Direct compression, 183, 323–324
Direct current, 369, 375
Discharge conveyor, 221
Discharge pressure, 361
Disk speed, 301–302
Disk, 298
Dispersive mixing, 33, 60, 78, 80, 92, 158–160, 269, 325, 335, 388
Displacement pump, 157
Dissolution rate, 250–251, 261–262, 264, 266–267, 269
Distributive mixing, 33, 78, 83, 88, 92, 160, 325, 335, 388
Dome extruder, 289–290
Downstack, 236
Drawdown, 211, 240
Drive system, 227
Drug loading, 198
Drug release, 187–188, 195, 200, 248
Dry granulation, 183, 323–324
Dry mixing, 281, 314
Drying, 303, 314
Dry-ingredient preparation, 117
Dumbbell, 300–302
Dust extraction hood, 347

Edge bead, 232, 234
Edge pinner, 234
Elasticity, 139
Electrical discharge machining, 109
Ellipsoid, 300
Emcompress, 193

End plate, 289
Entrained air, 163
Estradiol hemihydrate, 250
Ethylcellulose, 188, 193
Eudragit L, 190
Eudragit S, 189
Eutectic, 246, 265
Excipient, 187–188, 304
Explotab, 189
Extensional mixing, 74, 86
Extrudate, 184, 193, 205, 270, 284, 287–289, 295, 299
Extrusion/spheronization, 187, 277–281, 288, 297–298, 301–303, 314–316

Factory Acceptance Testing, 349
Feed hopper, 123
Feed rate, 156, 160, 361
Feed section, 46, 48–49, 186
Feeding equipment, 111, 155, 384, 387
Feedroll, 172
Feedstock, 43–44, 111
Film, 229
Flex lip, 238
Flight depth, 20
Flood feeding, 41, 111, 154, 160
Flow rate, 162
Flow-restriction device, 160
Fluid bed process, 303, 325–326, 330, 340
Fluted mixer, 62
Force displacement profile, 284, 293
Formulation factor, 191
Formulation variable, 303
Forwarding element, 78
Frame, 228
Free extrusion, 210
Free flowing, 41
Free volume, 20, 162
Fugitive dust, 347
Fume exhaust, 347
Funicular state, 284–285

Fusion, 184, 246, 265
Fusion-solvent method, 265

Gain-in-weight batching, 115
Gap, 159, 229
Gear pump, 162, 227
Gearbox, 350
Gearing, 19
Gear-type extruder, 292
Gelucire, 250
Glass solution, 246, 265
Glass suspension, 246, 265
Glass transition temperature, 199, 200,
    205, 248
Glatt granulator, 333
Glatt multicell, 328, 331
GMP extruder design, 29
Granulating fluid, 285–286, 288, 292,
    297, 299, 303, 309, 311, 314
Granulation, 282–283, 314, 323, 335,
    338–340
Gravimetric feeding, 121, 155–156
Gravity fed extruder, 288, 291, 295,
    297
Griseofulvin, 245, 262
Gris-PEG, 246
Grooved feed section, 53

Hammer mill, 263
Heat energy, 142
Heat exchanger, 367
Heat transfer fluid, 363
Heat transfer, 70, 92
Heating, 363
Heinen Continuous Fluid Bed, 331
Hexalobal, 97
High shear granulator, 326
High-energy mixer, 288
High-pressure barrel feed section, 52
High-shear granulation, 191, 327,
    329–330
High-shear mixer, 325, 327
Holdover time, 314

Homogenization, 263
Hopper, 41, 228
Hot-melt extrusion, 184
Human/machine interface (HMI), 377,
    387
Hydrochlorothiazide, 304, 337
Hydrocortisone, 255
Hydrolysis, 253
Hydroxypropylcellulose, 190, 193
Hydroxypropylmethylcellulose phthalate,
    189

Ibuprofen, 255
Implant, 255
Indomethacin, 251
Injection molding, 249, 388, 390
Inside diameter, 20
Installance Qualification, 348–349
Installation, 345–346
Integral band, 369
Intermesh gap, 20
Intermeshing twin screw, 8–10, 13, 15,
    162–163
Itraconazole, 255
Iverson mixer, 330, 334–335

Kneading disc, 83, 158, 336
Knoll Soliq, 388

L/D ratio (see Length-to-diameter ratio)
Lacidipine, 251
Lactose, 193
Ladder logic language, 377
Lamination, 232
Land length, 162
Length-to-diameter ratio, 20, 40, 163
Lidocaine HC1, 255
Line gap adjustment, 238
Line speed, 226
Lip opening, 229
Liquid feeding, 35, 118
Load cell, 121, 132
Lobal pool, 86

Loss-in-weight batching, 117
Loss-in-weight controller, 126
Loss-in-weight feeding, 116, 121, 157
Loviride, 255
Low-shear granulator, 326

Machine direction orientation, MDO, 230
Mandrel, 103
Manifold, 102
Marumerizer, 277
Mass flow, 116, 118, 123–124, 132, 157
Mass transfer, 70, 388
Material property, 304
Materials of construction, 350
Matrix pellet, 200
Matrix tablet, 183, 190, 195–196, 201, 205, 249
Mechanical energy, 92, 157
Melt extrusion, 2, 19, 153, 161, 184, 191, 249, 251, 253–254, 390
Melt granulation, 327
Melt homogenization, 65
Melt pressure, 355, 361, 378
Melt pump, 162
Melt seal, 161
Melt separation screw, 56
Melt temperature, 139, 164–165, 361, 380
Melt thermocouple, 142
Melt viscosity, 150, 159–160, 162
Melting zone, 58, 186
Melting, 55
Metallurgy, 385
Metering zone, 186
Microcompounding, 148, 270
Microcrystalline cellulose, 193, 279
Micro-feeder, 384
Micropelletization, 35, 179
Microwave energy, 325
Mixing bowl, 267–268
Mixing element, 59, 78

Mixing flow, 99
Mixing pin, 60
Mixing time, 285
Molecular solid dispersion, 247–248, 252, 265
Monitoring device, 185
Monitoring, 377
Morphology development, 159
Motor load, 355, 361, 378
Motor rotation, 354
Motor speed, 361
Multiparticulate, 162, 278, 312

NanoSystem, 264
Neckdown, 229
Nicardipine hydrochloride, 250
Nickel-based alloy, 385
Nip, 289–290, 292
Nip force, 236
Nip roll, 228
Niro granular, 332
Non-drag-flow pumping, 96
Nonintermeshing twin screw extruder, 8, 10, 71

On-demand cutting, 219
Operational qualification, 348, 353
Operator interface, 375
Ophthalmics insert, 255
Optical microscopy, 268
Orbiting screw mixer, 326
Oswald ripening, 264
Output, 40
Outside diameter, 20
Overflight gap, 20
Oxidation, 194, 253

Particle agglomeration, 264
Particulate dispersion, 265–266, 268, 270, 274
Particulate, 269
Patent, 5, 10, 13, 15
PC-based control, 377

Pellet, 171, 184, 188, 277, 279–280, 286
Pellet uniformity, 282
Pelletization, 171
Percolation theory, 200
Percolation threshold, 200
Performance qualification, 348, 356
Phenylpropanolamine HCI, 191
Planetary mixer, 326
Plasticizer, 188, 194
PLC-based control (*see* Programmable logic controller)
PLI (*see* Pounds per linear inch)
Poly (vinyl acetate), 199
Polyethylene glycol, 188
Polyethylene-co-vinyl acetate, 188
Polymeric matrix, 253
Polyvinylalcohol, 264
Polyvinylpyrrolidone, 264
Porosity, 307, 310
Pounds per linear inch, 236, 237
Powder x-ray diffraction, 268
Powder x-ray diffractogram, 272
Precipitation, 265
Precirol, 193
Precision drive, 387
Premix, 112
Pressure, 148
Pressure drop, 106, 161–162
Pressure gradient, 150
Pressure variation, 57–58
Preventive maintenance, 352
Process characterization, 139, 142
Process flow chart, 281
Process variable, 292, 338
Product thickness, 226
Product tolerance, 226
Product width, 226
Production capacity, 164
Profile, 209
Programmable logic controller, 376–377
Proportional band, 369
Proportional-integral-derivative (P-I-D) algorithm, 368

Pull roll, 230
Pulling device, 210
Pumping, 57
Purging material, 31

Qualification, 346
Quinidine sulfate, 304

Radial extruder, 290, 291
Radial pattern, 298
Ram extruder, 1, 96, 99, 184, 283–284, 288, 293
Rate-controlling agent, 191
Readout, 39
Recipe, 111
Refill, 112, 123, 125–127, 131
Relative viscosity, 142
Release profile, 188, 304
Release rate, 304, 307
Residence time distribution, 105
Residence time, 20, 105, 142–143, 147, 156, 158, 161, 164–165, 253, 291, 301–302, 314
Residual moisture, 161, 163
Resistance temperature detector, 366
Response parameter, 352
Restrictive screw element, 161
Restrictor, 102
Rheograph, 139, 143
Rheological behavior, 162
Rheometry testing, 138
Ribbon blender, 326
Roll design, 231
Roll diameter, 231
Roll extruder, 291
Roll feeder, 51
Roll gap, 241
Rolling bank, 242
Rotary knife cutter, 218
Rotor, 172
Rotor blade, 141–142

Safety, 353
Sanitation stress-rate threshold, 76
Scale-up, 33, 91, 93, 314, 390
Scanning electron microscopy, 202, 205, 301, 340
Screen perforation, 296
Screw channel, 42
Screw design, 20, 159, 337
Screw diameter, 40
Screw element, 77
Screw feeder, 118, 155
Screw speed, 155, 159–160, 378
Screw-fed extruder, 288, 291
Segmented screw, 22, 32
Self-cleaning, (*see* Self-wiping)
Self-wiping, 27, 72–73
Serial communication, 374
Shape extrusion, 209
Shark skinning, 283, 293, 296, 311
Shear, 88, 142
Shear cutter, 217
Shear energy, 142
Shear heating, 142
Shear mixing, 89
Shear rate, 74, 142, 150, 159–160
Shear sensitivity, 139
Shear stress, 89, 150, 158–159
Shear thinning, 89
Sheet die, 161, 237
Sheet extrusion, 235
Sheet slitter, 239
Shuttle system, 316
Side fed die, 104
Side feeder, 338
Sigma blade mixer, 326
Simple eutectic mixture, 246
Simulation, 137, 150
Single-screw extruder, 1, 4–6, 24, 39, 137, 150, 154, 383–384
Slitter station, 232
Slotted flight mixer, 61
Sodium chloride, 190
Sodium lauryl sulfate (SLS), 268

Solid dispersion, 246, 249, 254, 265
Solid solution, 246, 265
Solids transportation, 41–45
Solvent, 265
Specific energy, 24, 158, 164–165
Specific thermal energy, 157–158
Specification qualification, 349
Speed control, 369
Speed regulation, 371
SPEX 6800 freezer mill, 270
Spheronization (*see* Extrusion/Spheronization)
Spheronizer, 277, 288, 299–300
Spiral mandrel die, 104
Spironolactone, 245
Spreader roll, 235
Stability, 188, 194, 248, 254
Stabilizing cage, 228
Starve-fed, 60, 111, 155, 160
Statistical experimental design, 280
Steady state, 337
Sterotex, 193
Strain rate, 74
Strand breakage, 174
Strand pelletizer, 172
Stress rate, 75
Stress-rate profile, 91
Subprocess, 94
Substrate, 233–234
Sucroester, 250
Sulfadiazine, 262
Sulphathiazole, 246
Supercritical fluid, 263, 389
Surface coating, 108
Surface morphology, 202
Surface tension, 311
Surging, 44, 155
Sustained-release, 200
S-wrap, 227

Takeoff device, 215
Tandem extruder, 95
Tandem screw, 58

Tangential, 10
TEFC motor, 371
Temperature control, 46, 240, 362, 368
Temperature sensitivity, 139
Temperature, 148, 161, 361
TENV motor, 371
Tetracycline, 262
Theophylline, 200, 304, 312
Thermal degradation, 188
Thermal energy, 157, 165
Thermal gradient, 157
Thermocouple, 366–367
Throughput rate, 226
Tolbutamide, 251
Topical film, 255
Torque, 135, 136, 139, 148, 284–285, 337
Torque curve, 143
Torque rheometer, 135, 137–138, 140–141, 144–145, 147, 150, 283–284, 313
Torque signal, 145
Torque transducer, 137
Torque-limiting coupling, 355
Transdermal, 100, 161, 188, 226
Traveling saw, 217
Tray dryer, 303
Triacetin, 188
Troglitazone, 246
Troubleshooting, 23, 166
Twin screw extruder, 1, 6–7, 20, 26, 69, 71, 73, 85, 88, 137, 145, 153, 155, 286, 288, 325, 334, 383, 385

Underwater pelletizing, 176
Unit operation, 70, 73, 76, 93
Unwind shaft, 233
Unwind station, 232
Upstack, 236
Urethathane, 34
User requirements specification, 349

Vacuum calibration, 213
Vacuum level, 161, 361
Vacuum pump, 351
Validation, 345, 348–349
Velocity gradient, 159
Vent flooding, 58
Ventilation nozzle, 128
Venting, 58, 161
Vertical screw, 51
Vibrating Ring Droppo Pelletizer, 178
Vibratory feeder, 155
Vinylpyrrolidone-vinylacetate copolymer, 250
Viscoelasticity, 312
Viscosity, 75, 139, 142, 150, 159
Viscous dissipation, 157–158
Viscous heating, 74, 89
Volume, 92
Volumetric feeding, 113, 121, 156, 166–167
Vortec tube, 174

Water bath, 172
Water ring pelletizer, 177
Waterslide pelletizing, 173
Wax, 189–193
Web guiding, 239
Web path, 226
Wet granulation, 183, 323–325, 327, 329, 334–335, 390
Wet massing, 324
Wettability, 267
Wetting agent, 264
Wheel puller, 215
Winder, 228, 235

X-ray powder diffraction, 250

Zetachron, 388
Zoning element, 83